Materials Processing Under the Influence of External Fields

MATERIALS PROCESSING
UNDER THE INFLUENCE OF EXTERNAL FIELDS

TMS Member Price: $42   TMS Student Member Price: $33
                    List Price: $60

Related Titles

- *Solidification Processing of Metal Matrix Composites,*
  edited by N. Gupta and W.H. Hunt
- *Surface Engineering in Materials Science III,*
  edited by Arvind Agarwal, S. Seal and N.B. Dahotre

HOW TO ORDER PUBLICATIONS

For a complete listing of TMS publications,
contact TMS at (800) 759-4TMS or visit the TMS Document Center at
http://doc.tms.org:

- Purchase publications conveniently online.
- View complete descriptions, tables of contents and sample pages.
- Find award-winning landmark papers and reissued out-of-print titles.
- Compile customized publications that meet your unique needs.

MEMBER DISCOUNTS

TMS members receive a 30% discount on TMS publications. In addition, members receive a free subscription to the monthly technical journal *JOM* (both in print and online), discounts on meeting registrations, and additional online resources to name a few of the benefits. To begin saving immediately on TMS publications, complete a membership application when placing your order in the TMS Document Center at http://doc.tms.org or contact TMS.

Telephone: (724) 776-9000 / (800) 759-4TMS
E-mail: membership@tms.org or publications@tms.org
Web: www.tms.org

Materials Processing Under the Influence of External Fields

Proceedings of a symposia sponsored by
TMS (The Minerals, Metals & Materials Society)

TMS 2007 Annual Meeting & Exhibition
Orlando, Florida, USA
February 25-March 1, 2007

Edited by

Qingyou Han
Gerard Ludtka
Qijie Zhai

A Publication of

A Publication of **The Minerals, Metals & Materials Society (TMS)**
184 Thorn Hill Road
Warrendale, Pennsylvania 15086-7528
(724) 776-9000

Visit the TMS Web site at
http://www.tms.org

Library of Congress Catalog Number 2007921347
ISBN 978-0-87339-664-6

If you are interested in purchasing a copy of this book, or if you would like to receive the latest TMS publications catalog, please telephone (724) 776-9000, ext. 270, or (800) 759-4TMS.

MATERIALS PROCESSING UNDER THE INFLUENCE OF EXTERNAL FIELDS
TABLE OF CONTENTS

Materials Processing under the Influence of External Fields

Session I

Session II

Session III

Session IV

Session V

PREFACE

These proceedings represent the written record of the symposium "Materials Processing Under the Influence of External Fields," which was held as part of the TMS 2007 Annual Meeting & Exhibition in Orlando, Florida, United States, February 25–March 1. The symposium was co-sponsored by the Solidification Committee, the Aluminum Committee, and the Magnesium Committee, of the Materials Processing & Manufacturing Division (MPMD) of The Minerals, Metals & Materials Society (TMS).

The intent of this symposium was to review the recent advances in the area of materials processing under the influence of external fields, to integrate the perspectives from academia, national laboratories and private corporations, and to contribute to the transfer of science and technology among these three types of institutions, and to link potential solutions to current challenges. Topics included the use of magnetic field, electro-magneto field, electric field, ultrasonic vibrations, and microwaves for the processing of metal and alloys, organic materials and ceramics materials. These proceedings indicate that the development of the next generation of high performance materials may be enabled through processing in extreme environments involving various field phenomena.

The manuscripts submitted have been independently peer reviewed. These papers represent significant contributions on a wide variety of fundamental and technological aspects. The authors present physical-based mathematical models, theories, simulations, and new and innovative techniques and processes, and demonstrate the potential for applications in materials processing.

We thank the authors for providing their manuscripts in a timely fashion that has allowed these proceedings to be available at the symposium. It is with pleasure that we thank the MPMD council for supporting the publication of these proceedings, and members of the TMS staff for their enthusiastic and dedicated assistance.

Qingyou Han, Oak Ridge National Laboratory

Gerard Ludtka, Oak Ridge National Laboratory

Qijie Zhai, Shanghai University

Materials Processing Under the Influence of External Fields

Session I

Materials Processing under the Influence of External Fields
Edited by Qingyou Han, Gerard Ludtka, Qijie Zhai
TMS (The Minerals, Metals & Materials Society), 2007

TIME-RESOLVED ANALYSES OF MICROSTRUCTURE IN ADVANCED MATERIALS UNDER MAGNETIC FIELDS AT ELEVATED TEMPERATURES USING NEUTRONS

Gerard M. Ludtka[1], Frank Klose[2], Roger Kisner[3], Jaime Fernandez-Baca[1],
Gail Mackiewicz-Ludtka[1], John Wilgen[3], Roger Jaramillo[1], Louis Santodonato[2],
Xun-Li Wang[2], Camden R. Hubbard[1], Fei Tang[1]

[1]Materials Science & Technology
[2]Spallation Neutron Source
[3]Engineering Science and Technology
Oak Ridge National Laboratory
1 Bethel Valley Rd, Oak Ridge, Tennessee, 37831, USA

Keywords: Time-Resolved, Microstructure, Materials, Magnetic Fields, Magnetic Processing, Neutron Science, Neutron Scattering, Temperature, Phase Transformations, Ferrous Alloys, In-situ Analyses, Quantitative

Abstract

Fundamental science breakthroughs are being facilitated by high magnetic field studies in a broad spectrum of research disciplines. Furthermore, processing of materials under high magnetic fields is a novel technique with very high science and technological potential. However, currently the capability does not exist to do in-situ time-resolved quantitative analyses at high magnetic field strengths and elevated temperatures. Therefore, most measurements are performed ex situ and do not capture the microstructural evolution of the samples during high field exposure. To address this deficiency, we are developing high field magnet processing and analyses systems at the High Flux Isotope Reactor and the Spallation Neutron Source at the Oak Ridge National Laboratory which will link the analytical capabilities inherent in neutron science to the needs of magnetic processing research. Our goal is to apply advanced neutron scattering techniques to explore time-resolved characterizations of magnetically driven alloy phase transformations under transient conditions. This paper will provide an overview of the current status of this research endeavor with preliminary results obtained on ferrous alloys.

Introduction

High magnetic field processing is proving to be an innovative and revolutionary new focus area [1-12] that is creating the basis for an entirely new research opportunity for materials and materials process development. Based on fundamental thermodynamic principles, our experimental and modeling efforts over the last several years [6-14] have clearly demonstrated that phase stability (conventional phase diagrams) can be dramatically altered through the application of an ultrahigh magnetic field. The ramifications of this research include the existence of entirely new phases, magnetically induced isothermal phase transformation reversal, bulk nanocrystalline materials, texture enhancement, reduced residual stress, carbon nanotube formation, advanced superconducting materials, and novel microstructures with enhanced performance. More broadly, fundamental science breakthroughs are being facilitated by high magnetic field studies in a broad spectrum of research disciplines such as condensed matter

3

physics, materials, biology, and medicine. Therefore, incorporating in-situ neutron scattering characterization methods with high magnetic field research has exceptional potential.

Our ex-situ research [6-14] clearly shows the breakthrough science that can be accomplished using ultrahigh magnetic fields. However, consistent throughout this research (and throughout the refereed literature) is the fact that all microstructural analyses and modeling validation were conducted *after* the magnetic processing experiments were concluded. Therefore the influence of magnetic fields on phase equilibria existing during experiments was characterized on specimens without a magnetic field superimposed. There exists the strong possibility that phase equilibria shifting could have occurred while removing the magnetic field, even at ambient temperatures. This potential discrepancy can be significant when assessing the validity of calculations (e.g., the ab-initio Local Spin Density modeling endeavor of reference 7) which are critical to accurately predicting magnetically enhanced phase diagrams. Accurate, validated predictions are essential to propose future experiments and materials development efforts. Therefore, for our research we recognized the critical need to establish a robust ultrahigh magnetic field environmental system for use at both the upgraded Oak Ridge National Laboratory High Flux Isotope Reactor (HFIR) and recently commissioned Spallation Neutron Source (SNS) for conducting advanced time-resolved characterization studies. Such new capabilities would facilitate developing the fundamental science necessary to achieve future major breakthroughs based on magnetic field effects on materials across many disciplines.

Design, fabrication and test of sample environment system

Initially, neutron scattering measurements will be performed at the HFIR using a 5 Tesla superconducting magnet. The challenge was to develop a sample insert that could provide sustained sample temperatures of 800-1000°C while limiting the heat load to the magnet cryostat to less than 10 watts. A sample insert was developed that provides the capability for inductively heating steel samples to 1000°C while limiting the heat load to the magnet cryostat. To minimize deleterious interactions of the neutron beam with the sample insert, an air-cooled (rather than water-cooled) induction coil is used to heat the sample. High temperature thermal insulation is provided by low-density alumina fiber insulation. To facilitate the process of changing out samples, the thermally insulated steel samples are enclosed in a quartz tube that can readily be withdrawn from the magnet. Each sample tube includes a type-S thermocouple, and an inert purge gas manifold. A photo showing a quartz sample tube containing an insulated sample positioned within the aluminum induction heating coil is shown in Figure 1.

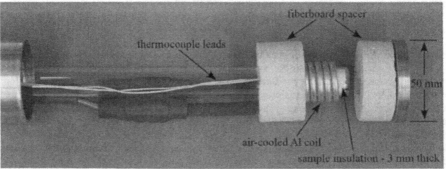

Figure 1. Photograph of specimen holder and key components.

After fabrication and assembly of the sample insert was completed, a 52100 steel specimen was given a heat treatment to confirm temperature control as well as heating and cooling requirements for sample temperatures in the 800-1000°C range. An S-type thermocouple was attached to the specimen and provided feedback control to the induction heating unit. The results of this test are shown in Figure 2. The figure plots the recorded temperature throughout the applied thermal cycle. The system was programmed to heat the 12 mm long by 8 mm diameter specimen to 1000°C in two minutes, hold for three minutes and cool to 750°C in two minutes followed by a six minute hold and air cool. The figure confirms the functionality of the insert and displays good temperature control.

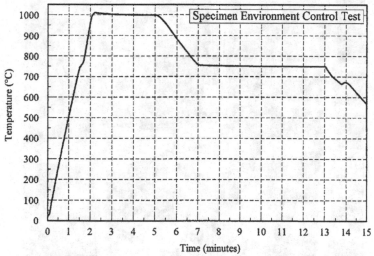

Figure 2. Temperature measurements displaying heating, cooling and temperature control for environment insert system.

Additionally, the sample insert was test-fitted into the 5 Tesla magnet to verify mechanical compatibility. A top view of the fully assembled system installed in the magnet cryostat is shown in the photo in Figure 3. Development of the high temperature sample insert was completed as planned and is currently available for deployment on the neutron beam line at HFIR. A data acquisition and induction heating control system will be employed on the experimental apparatus that will be used for neutron scattering under a high magnetic field at HFIR. The function of the system is to control heat-up, maintain steady-state heated conditions, and cool experimental samples.

Preliminary neutron scattering experiments

Preliminary neutron scattering experiments were performed at the HFIR using steel specimens. The experiments were conducted at room temperature, without high magnetic fields and used plated and non-plated specimens. The plating consisted of a 0.6 mm layer of copper. The goal of these experiments was to develop neutron scattering data analysis and characterization methods using the Wide Angle Neutron Diffraction (WAND) and Neutron Residual Stress Facility 2 (NRSF2) instruments. These two instruments compliment each other well as the NRSF2 provides high precision d-spacing measurements for determination of crystallographic lattice parameters and the WAND can obtain the full diffraction pattern and apply this information in a Rietveld analysis to determine phase weight fractions.

Figure 3. Heating/cooling insert shown installed on 5 Tesla superconducting magnet at HFIR.

Two steel alloys, a medium carbon 1045 steel and a high carbon, high chromium 52100 steel were probed and diffraction patterns were analyzed to determine crystal lattice parameters and phase fractions. For the sake of brevity, only results for non plated specimens and WAND data will be presented here. Table I shows lattice parameter measurements for ferrite (BCC) in both steels. The results are compared to X-ray diffraction measurements for 51XX series steel [15].

A diffraction pattern for an uncoated 52100 specimen is shown in Figure 4. A Rietveld analysis has been performed for estimating phase weight fractions. This particular analysis estimates 17% cementite (Fe_3C) and 83% ferrite (BCC). This result is reasonable although the cementite content is exaggerated compared to ThermoCalc predictions of ~15 wt.% cementite. These apparently high values for cementite fraction were consistent for 1045 results and suggest that some

6

additional refinement may be required in the analysis. This accomplishment provided scoping data for developing and refining diffraction pattern analysis methodology and lays the foundation for later experiments.

Table 1. Ferrite lattice parameter measurements in angstroms.

Steel alloy	WAND ($\lambda = 0.1476$nm, $2\theta_o = -0.24°$)	X-ray measurements (Cavin et al)
1045	2.8659 ± 0.00002	2.8649
52100	2.8682 ± 0.00003	

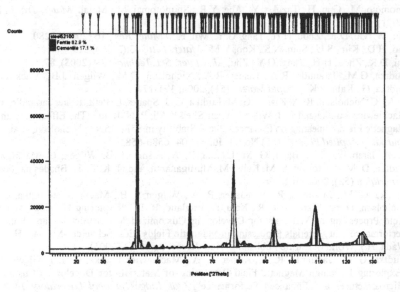

Figure 4. WAND diffraction pattern and applied Rietveld analysis for uncoated 52100 specimen.

Future Facility Development Activities

The next step in the project is to combine the elevated temperature capability with a 5 Tesla magnetic field and perform neutron scattering experiments. For the current HFIR operations schedule, these experiments should be conducted in the first half of 2007. High purity, levitation cast Fe-C alloys have been obtained and will be used in these experiments. The goal of these experiments is to obtain precise information regarding shifts in the Fe-C phase diagram due to the high magnetic field.

A major advancement of the program will be the addition of superconducting magnets with field strength much greater than 5 Tesla. One of the additions is a 16 Tesla superconducting magnet system at SNS (being built now and scheduled for commissioning in Spring 2009). A deliverable of our project is to provide an environment insert similar to the one described in this paper for the 16 Tesla magnet. The second addition is associated with the planned Zeemans high magnetic field facility at the SNS. For this program, specifications and designs are currently being written for a ~40 Tesla hybrid magnet. Such a magnet will merge the state-of-the-arts in magnet technology and neutron scattering. It is anticipated that an elevated temperature insert will be developed for this advanced system.

Acknowledgements

This research is sponsored by the Laboratory Directed Research and Development Program of Oak Ridge National Laboratory (ORNL), managed by UT-Battelle, LLC, for the U.S. Department of Energy under contract number DE-AC05-00OR22725.

References

1. Choi, J-K., Ohtsuka, H., Xu, Y., Choo, W-Y. *Scripta Mater.*, 43 (2000), 221.
2. Enomoto, M., Guo, H., Tazuke, Y., Abe, Y.R., Shimotomai, M., Metall. *Mater. Trans.*, 32A (2001), 445.
3. Liu, W., Ou, D.R., Zhou, H.H., Tang, G.Y., Wu, F., *J Mater Res.*, 16 (2001), 2280.
4. Joo, H.D., Kim, S.U., Shin, N.S., Koo, Y.M., *Mater. Lett.*, 43 (2000), 225.
5. Ou, D.R., Zhou, H.H., Tang, G.Y., Zhu, , *J., Mater. Sci. Technol.*, 19 (2003), 87.
6. Ludtka, G.M., Jaramillo, R.A., Kisner, R.A., Nicholson, D.M., Wilgen, J.B., Mackiewicz-Ludtka, G., Kalu, P.K., *Scripta Mater.*, (51), 2004, 171-174.
7. D. M. C. Nicholson, R. A. Kisner, G. M. Ludtka, C. J. Sparks, L. Petit, Roger Jaramillo, G. Mackiewicz-Ludtka and J. B. Wilgen, Askar Sheikh-Ali, P. N. Kalu[2] "The Effect of High Magnetic Field Annealing on Enhanced Phase Stability in 85Fe-15Ni", Nicholson, et al, *Journal of Applied Physics*, (95) No. 11, June 2004, 6580-6582.
8. R. A. Jaramillo, S. S. Babu, G. M. Ludtka, R. A. Kisner, J. B. Wilgen, G. Mackiewicz-Ludtka, D. M. Nicholson, S. M. Kelly, M. Murugananth, and H. K. D. H. Bhadeshia, *Scripta Materialia*, (52), 2005, 461-466 .
9. Ludtka, G. M., Jaramillo, R. A., Kisner, R. A., Wilgen, J. B., Mackiewicz-Ludtka, g. M., Nicholson, D. N., Watkins, T. R., Kalu, P., England, R. D., "Exploring Ultrahigh Magnetic Field Processing of Materials for Developing Customized Microstructures and Enhanced Performance", in Materials Processing in Magnetic Fields (Eds. Schneider-Muntau, H. J. and Wada, H.), World Scientific Publishing, New Jersey, pp 55-65 (2005).
10. Ludtka, G. M., Jaramillo, R. A., Kisner, R. A., Mackiewicz-Ludtka, G. and Wilgen, J. B., "Exploring Ultrahigh Magnetic Field Processing of Materials for Developing Customized Microstructures and Enhanced Performance", *Oak Ridge National Laboratory Technical Report* ORNL/TM-2005/79, April 2005.
11. Jaramillo, R.A., Babu, S., Miller, M.K., Ludtka, G.M., Mackiewicz-Ludtka, G. Kisner, R.A., Wilgen, J.B. and Bhadeshia, H.K.D.H, "Investigation of Austenite Decompositions in high-carbon high-strength Fe-C-Si-Mn steel under 30 Tesla Magnetic Field", *Proceedings of the TMS Conference "Solid-Solid Phase Transformations in Inorganic Materials 2005"*, May 29-June 3, 2005, Phoenix, AZ.
12. Jaramillo, R.A., Ludtka, G.M., Kisner, R.A., Nicholson, D.M., Wilgen, J.B., Mackiewicz-Ludtka, G., Bembridge, N. and Kalu, P.N., "Investigation of Phase Transformation Kinetics and Microstructural Evolution in 1045 and 52100 Steel under Large Magnetic Fields", *Proceedings of the TMS Conference "Solid-Solid Phase Transformations in Inorganic Materials 2005"*, May 29-June 3, 2005, Phoenix, AZ.
13. J. B. Wilgen, R. A. Kisner, Gerard. M. Ludtka, Gail M. Ludtka, and R. A. Jaramillo, "Method for Non-contact Ultrasonic Treatment of Metals in a High Magnetic Field", Department of Energy Invention Disclosure No. S-105,133, submitted April 6, 2005.
14. R. A. Kisner, J. Wilgen, G. M. Ludtka, R. A. Jaramillo, and Gail Mackiewicz-Ludtka, "Thermal and High Magnetic Field Treatment of Materials and Associated Apparatus" ,U.S. Patent Application No. 11/109,376 filed 4/19/2005.
15. O.B. Cavin, D.A. Carpenter and C.R. Hubbard, ORNL unpublished data.

ISOTHERMAL PHASE TRANSFORMATION CYCLING IN STEEL BY APPLICATION OF A HIGH MAGNETIC FIELD

Gerard M. Ludtka[1], Roger A. Jaramillo[1], Gail Mackiewicz-Ludtka[1], Roger A. Kisner[2], John B. Wilgen[2]

[1] Materials Science & Technology, Oak Ridge National Laboratory, Oak Ridge, TN 37831
[2] Engineering Science & Technology, Oak Ridge National Laboratory, Oak Ridge, TN 37831

Keywords: magnetic field cycling, 5160 steel, transformation reversal, isothermal transformations, austenite, ferrite, spheriodized cementite

Abstract

A phase transformation reversal via the application and removal of a large magnetic field was investigated. Because a large magnetic field can alter the phase equilibrium between paramagnetic austenite and ferromagnetic ferrite, volume fractions for each phase constituent can be modified at constant temperature by changing the magnetic field strength. In this research elevated temperature isothermal hold experiments were performed for 5160 steel. During the isothermal hold, the magnetic field was cycled between 0 and 30 Tesla. As companion experiments, temperature cycling and isothermal holds were performed without magnetic fields. The resulting microstructures were examined using optical and SEM metallography. These microstructures indicate that a portion of the microstructure experiences isothermal transformation cycling between austenite and ferrite due to the application and removal of the 30T (Tesla) magnetic field.

Introduction

The application of a large magnetic field as a new tool for developing unique microstructures in ferrous materials is becoming realized as superconducting magnet technology for efficiently generating large (>10T) magnetic fields becomes feasible. It has been demonstrated that magnetic fields significantly affect the thermodynamics of phase equilibria and that new microstructures are possible for conventional alloy chemistries. Model predictions of this effect [1] and isothermal transformation experiments [2] have been reported. Data showing a significant increase in transformation temperature during continuous cooling for 1045 steel have been reported by the authors [3]. This result is significant in that in situ temperature measurements of thermal recalescence associated with austenite decomposition provide a relative shift in transformation temperature due the application of a 30T magnetic field. These results indicate a 3°C/T increase in transformation temperature. Additional research has shown that unique microstructures can be obtained via the application of a large magnetic field. A high carbon, high silicon steel exhibited very fine pearlite with a lamellar spacing of approximately 50 nm during continuous cooling with a large magnetic field [4].

In this work, the ability of a magnetic field to isothermally cycle the phase transformation between ferromagnetic ferrite and paramagnetic austenite was investigated. By holding the material at a temperature just above the austenite/ferrite phase boundary, the volume fractions of the phase could be modified through the application of a large magnetic field. Microstructural and hardness evaluations are presented that reveal significant modifications in microstructure associated with magnetic field cycling when compared to an isothermal hold without magnetic field treatment.

Theory

The magnetic field changes the free energy of the austenite-ferrite equilibrium such that the ferromagnetic ferrite is stabilized relative to the paramagnetic austenite. The magnetic field contribution to the change in free energy has been estimated to be 12.6 J/mol/T [5]. Applying this estimate, the effect of the magnetic field on the 5160 steel pseudo-binary is shown in Figure 1. The figure plots phase boundaries for 5160 steel as calculated using ThermoCalc Fe-DATA database [6]. The black lines represent phase boundaries without the magnetic field effect; red lines are shifted phase boundaries assuming that the free energy of the ferrite is changed by -378 J/mol due to the application of a 30T magnetic field. This estimate shows a significant increase in Ae1 and Ae3 temperatures due to the magnetic field. The boundary shift suggests that for certain temperatures, the $\gamma \leftrightarrow \alpha$ + Fe$_3$C transformation can be fully reversed by cycling the magnetic field between 0 and 30T. Our experiments were designed to test this theory.

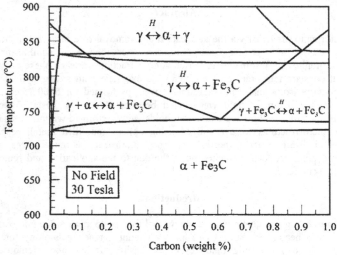

Figure 1. Calculated pseudobinary phase diagrams for a nominal 5160 steel chemistry. Black lines represent boundaries at 0 Tesla and red lines are boundaries with the application of a 30 Tesla magnetic field.

Experimental

Hollow cylindrical specimens with a 6-mm outer diameter, 1-mm wall thickness and 8-mm length were machined from a medium carbon 5160 steel. 5160 steel has a nominal composition of 0.80 wt.% Cr, 0.88 wt.% Mn and 0.60 wt.% C. Experiments were performed using a 32mm diameter bore resistive magnet with a 33T maximum field strength at the National High Magnetic Field Laboratory (NHMFL). Specimen heating and cooling was controlled using a custom designed induction heating coil coupled with a gas purge/quench system. The apparatus locates the specimen in the center of the bore (mid length) and can heat steel specimens up to 1100°C and maintain the high temperature for extended periods. Temperature control was obtained using a type "S" (Pt-10%Rh) thermocouple spot-welded on the outer diameter at the mid-length of the specimen. The system allows for the entire thermal cycle or any portion of it to be exposed to a high magnetic field.

A schematic of the thermal-magnetic cycles used in the experiments is shown in Figure 2. The figure plots temperature and magnetic field strength as a function of time. The temperature is ramped to 745°C in two minutes and maintained until a prescribed quench at step 1, 2 or 3 as shown in the figure. During the isothermal hold the magnetic field was ramped up to 30T and held for 10 minutes including the ramp up time. Each additional step is 10 minutes with the magnetic field cycled between 0 and 30T. To capture the evolution of the microstructure, an experiment was ended at the end of each 10 minute period and identified as "Step1," "Step 2," and "Step 3." Also, 40 minute isothermal hold without any magnetic field treatment was performed. A helium gas quench was applied at the end of each experiment.

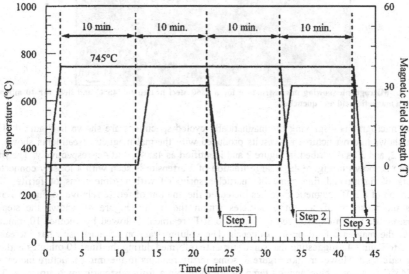

Figure 2. Schematic illustrating the temperature and magnetic field histories used for magnetic field cycling experiments. Black plot is magnetic field strength and Red (lighter) lines are temperature.

Results and Discussion

As a basis for comparison with magnetic field cycling experiments, a specimen was heat treated with an isothermal hold at 745°C for 40 minutes without any magnetic field treatment. This is the same thermal treatment represented as step 3 in Figure 2. The resulting microstructure is shown in Figure 3. The microstructure is predominantly martensitic (light gray) with isolated regions of ferrite (white). The presence of ferrite indicates that the material did not fully transform to austenite at 745°C and was a mixture of austenite and ferrite. During the quench, the austenite transforms into martensite and the ferrite remains constant. However, the volume fraction of ferrite is small and the application of a magnetic field should increase the amount of ferrite, therefore enriching the austenite in carbon and/or forming spherical cementite carbides.

11

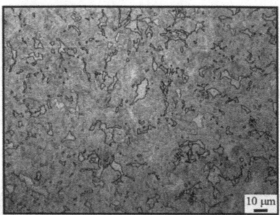

Figure 3. **Micrograph showing microstructure for a 5160 steel heated to 745°C and held for 40 minutes without a magnetic field and quenched.**

Optical micrographs displaying the magnetically cycled specimens are shown in Figure 4. The figure shows microstructures for results produced with thermal-magnetic treatments shown in steps 1, 2, and 3 as described in Figure 2 and identified as 4a, 4b, and 4c, respectively. The first 30T step is shown in Figure 4a, a large fraction of martensite is seen with a loosely connected network of ferrite and fine cementite particles sprinkled within some areas of ferrite. The application of a 30T magnetic field has increased the amount of ferrite relative to Figure 3 and shows the presence of cementite particles within the ferrite. Figure 4b shows the step 2 microstructure which experienced a 10 minute 30T treatment followed by another 10 minutes without the magnetic field prior to quench. The volume fraction of martensite has increased relative to Figure 4a suggesting that some austenite reformed during the final 10 minutes without a magnetic field. However, like Figure 3, some ferrite remains in the microstructure indicating that 745°C was not a high enough temperature to obtain a fully austenitic microstructure. The final micrograph in the series (Figure 4c) shows the microstructure after a final 10 minutes with a second 30T treatment. The microstructure is substantially different from Figures 4a and 4b. As expected the volume fraction of martensite has again decreased due to the isothermal transformation of austenite to ferrite and cementite. Additionally, the second 30T treatment step appears to have increased the amount of austenite transformed isothermally compared to the amount indicated for the first 30T treatment shown in Figure 4a. The amount of ferrite and spheroidized cementite provide strong evidence of isothermal transformation due to the application of a 30T magnetic field.

The interpretation of the microstructures shown in Figures 3 and 4 are substantiated by microhardness measurements. Table 1 shows the results of microhardness measurements performed on the four specimens. A 500 gram load was used and the reported HV is the average of five indentations.

Table I. Microhardness measurements for magnetically cycled and isothermal-only treated specimens. (500 g load, 5 indentations)

	Step 1 (Fig 4a)	Step 2 (Fig 4b)	Step 3 (Fig 4c)	No Magnetic Field (Fig 3)
Hardness (HV)	219	382	226	451
Std. Dev.	45	127	72	133

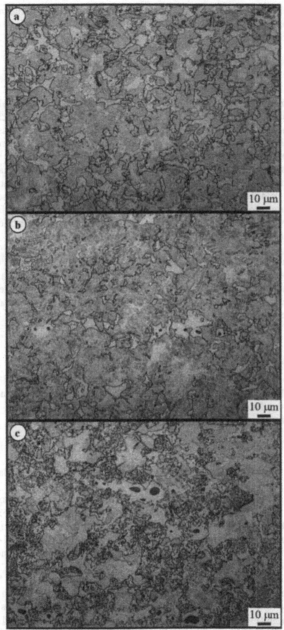

Figure 4. Optical micrographs of magnetically cycled 5160 steel. Microstructures in a), b) and c) correspond to steps 1, 2, and 3, respectively, from Figure 2.

The specimens with a magnetic treatment in the final step are softer than those without. This is due to the reduction in martensite associated with greater amounts of isothermal transformation during magnetic field treatment. The large standard deviations for the non-magnetically treated specimens would be expected given the limited number of indentations and the mixture of soft ferrite and hard martensite.

Summary and Conclusions

Magnetic field cycling experiments were performed to investigate isothermal transformation reversal via the application and removal of a large magnetic field. A 5160 steel was heated to 745°C and held while a 30T magnetic field was cycled. A specimen was quenched out at each step of the cycle and a simple isothermal treatment without a magnetic field was performed for comparison. The subsequent metallographic analysis and micros hardness measurements confirm the phenomenon of isothermal phase transformation reversal. Micrographs show modified amounts of martensite and ferrite/cementite due to the application of a 30T magnetic field. A large volume fraction of martensite is indicative of limited isothermal transformation of austenite during the 745°C hold and most transformation occurring during the subsequent quench. However, for the case when a magnetic field was applied in the final step, a reduced amount of martensite was observed and for the case of a second 30T treatment in the cycle, a well developed phase constituent of ferrite and spheroidized cementite was seen.

These results validate the use of a large magnetic field for phase transformation cycling. Such a processing tool could provide a means of microstructural refinement and accelerated spheroidization treatments. Also, the permeability of the magnetic field eliminates gradients that would typically occur when cycling a transformation via thermal cycle. These capabilities could promote the development of advance microstructures and components with superior performance.

Acknowledgements

We acknowledge Dr. Bruce Brandt and the staff at the National High Magnetic Field Laboratory for their support. Research sponsored by the Laboratory Directed Research and Development Program of Oak Ridge National Laboratory (ORNL), managed by UT-Battelle, LLC for the U. S. Department of Energy under Contract No. DE-AC05-00OR22725.

References

1. J-K Choi et al., "Effects of a strong magnetic field on the phase stability of plain carbon steels," *Scripta Materialia*, 43 (2000), 221-226.
2. M. Enomoto et al., "Influence of Magnetic Field on the Kinetics of Proeutectoid Ferrite Transformation in Iron Alloys," *Metall. Mater. Trans. A*, 32A (2001), 445-453.
3. G. M. Ludtka et al., "In situ evidence of enhanced transformation kinetics in a medium carbon steel due to a high magnetic field," *Scripta Materialia*, 51 (2004), 171-174.
4. R. A. Jaramillo et al., "Effect of 30T magnetic field on transformations in a novel bainitic steel," *Scripta Materialia*, 52 (2005), 461-466.
5. R.A. Jaramillo, G.M. Ludtka, R.A. Kisner, D.M. Micholson, J.B. Wilgen, G. Mackiewicz-Ludtka, N. Bembridge, and P.N. Kalu, "Investigation of phase transformation kinetics and microstructural evolution in 1045 and 52100 steel under large magnetic fields", in Solid-to-Solid Phase Transformations in Inorganic Materials 2005, Vol. 1: Diffusional Transformations, (Eds. J.M. Howe, D.E. Laughlin, J.K. Lee, U. Dahmen, W.A. Soffa), TMS, PA, pp. 893-898 (2005).
6. N. Saunders, "Fe-DATA, a database for thermodynamic calculations for Fealloys," Thermotech Ltd., Surrey Technology Centre, The Surrey Research Park, Guilford, Surrey GU2 7YG, U.K.

Materials Processing under the Influence of External Fields
Edited by Qingyou Han, Gerard Ludtka, Qijie Zhai
TMS (The Minerals, Metals & Materials Society), 2007

Solid state materials processing in a high magnetic field

Hideyuki Ohtsuka

National Institute for Materials Science
High Magnetic Field Research Station
3-13 Sakura, Tsukuba, Ibaraki 305-0003, Japan

Keywords: ferrite, carbon, alignment, diffusional transformation,
diffusion, transformation temperature

Abstract

Effects of a high magnetic field on diffusional transformation temperature have been investigated in pure iron, Fe-0.8C and Fe-25Co (mass%) alloys. It was found that the transformation temperature is increased with increasing applied magnetic field in all alloys and the experimentally determined data are not in good agreement with those calculated above 10T.

The increase of transformation temperatures in a magnetic field of 30T are 29.0°C, 31.2°C and

48.4°C for pure iron, Fe-0.8C and Fe-25Co alloys, respectively. No elongation or alignment of transformed structures was observed in these alloys. Effects of a high magnetic field on diffusion of carbon were also studied, and it was found that the diffusivity of carbon in ferrite is retarded in a magnetic field.

Introduction

With the rapid progress in the technology of superconducting magnet, a high magnetic field more than 10T becomes available in laboratory. As a result, processing in high magnetic field is emerging as an innovative method for controlling microstructure and properties of steels. Some magnetic field effects in steels have been found recently. For example, we reported that ferrite grains are elongated and aligned along the direction of magnetic field in austenite matrix for ferrite transformation in a high magnetic field [1-3]. Elongated and aligned structure was also found by reverse transformation from martensite to austenite [4,5]. Thermodynamic calculation shows phase equilibrium and Phase diagram of Fe-C system is changed in a high magnetic field [6]. In this report, we studied the effects of high magnetic field on thermodynamics and kinetics of phase transformation by the measurement of phase transformation temperature and carbon diffusion in Fe-C alloys.

Results and discussion

1. Phase transformation temperature

We have reported that ferrite transformation temperature from austenite to ferrite in pure Fe increases linearly (0.8K/Tesla) with magnetic field less than 10T [6]. For pearlite transformation in Fe-0.8C alloy, the transformation temperature increases about 1.5K/Tesla. It is interesting to find if the linear relationship still exists when the magnetic field is higher than 10 Tesla. Figure 1 shows the ferrite transformation (a) and pearlite transformation (b) temperatures change in a magnetic field up to 30 Tesla. The increased transformation temperature (ΔT) with magnetic field was compared between experimental results of this study and calculated value using molecular field theory, and shown in Fig. 1. For ferrite transformation in pure Fe, it was found that the calculated transformation temperature deviates from the linear relationship and increases faster in very high magnetic field. The experimentally determined transformation temperature is lower than that calculated using molecular field theory when the magnetic field strength is higher than 10 T. The ΔT calculated using molecular theory is almost the same as that calculated using

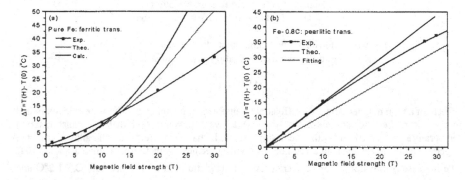

Fig.1 Phase transformation temperature in a high magnetic field. (a) Austenite to ferrite transformation in pure Fe and (b) austenite to pearlite transformation in an Fe-0.8C alloy.

susceptibility data that were measured in low field. It is interesting to note that when the magnetic field strength is higher than 15 T, the measured ΔT becomes much lower than that calculated using molecular field theory and using susceptibility data. For pearlite transformation in Fe-0.8C alloy (Fig. 1(b)), the measured ΔT values in magnetic field up to 30T are always higher than that calculated by Weiss theory. The ΔT increases linearly with magnetic field up to 10T and it deviates from the linear relationship when the field is higher than 10 T. Figure 1 shows that both susceptibility data measured in low field and molecular field theory are not suitable to calculate phase transformation temperature in high magnetic field. Simple linear relationship between transformation temperature and magnetic field does not exist. More theory and experimental works are needed to precisely predict phase transformation temperature in a high magnetic field.

Effects of a high magnetic field on austenite to ferrite transformation temperature in Fe-25Co alloys have been investigated. The transformation temperature increases with increasing applied magnetic field. The increment of transformation temperature is 48.4°C with applied magnetic field of 30T and this is larger than those of pure iron and Fe-0.8C because Fe-Co alloy has a larger magnetic moment than pure iron and Fe-0.8C. The increase of transformation temperatures in a magnetic field of 30T are 29.0°C, 31.2°C and 48.4°C for pure iron, Fe-0.8C and Fe-25Co alloys, respectively. In an Fe-25Co alloy transformed in a magnetic field of 10T, fairly large grains with many subgrain boundaries are observed but the elongated structure cannot be found in the specimen.

2. Carbon diffusion in austenite

Phase transformations (ferrite transformation, pearlite transformation, etc) at high temperature in steels are diffusive transformation. It is very important to know how the magnetic field affects atom diffusion in steels. Nakamichi et al [8] have found that the carbon diffusion in austenite is retarded by 40% in a 6 T magnetic field. Figure 2 shows the effect of high magnetic field on carbon diffusion in austenite and ferrite. The carbon diffusion experiments were carried out by diffusion couple method. First, a piece of pure Fe (15 x1 5 x 5 mm) and a piece of Fe-0.8C (15 x 15 x 5 mm) are pre-welded at 1000 ℃ for 30min in a vacuum furnace . Then diffusion experiment was carried at 930℃ for 5 hours without (a) or with (b) a magnetic field of 10 T followed by furnace cooling. The magnetic field is parallel to the carbon diffusion direction. At 930℃ both pure Fe and Fe-0.8C are in austenite state. So carbon atoms diffuse in a single phase from the high carbon side (Fe-0.8C, left) to low carbon side (pure Fe, right). During cooling, Fe-0.8C transformed to pearlite (dark) and pure Fe transformed to ferrite (bright). In the diffusion region, a mixture of pearlite and ferrite appeared. The amount of pearlite decreases and the amount of ferrite increases from left to right according to carbon concentration. A white line showed the original interface between Fe-0.8C and pure Fe. The effects of a high magnetic field on diffusivity of carbon in austenite state seems very small. Figure 3 shows another diffusion

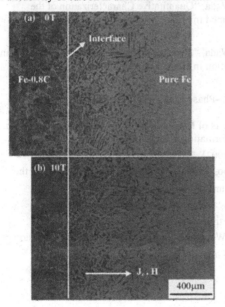
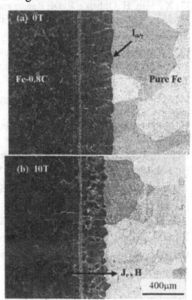

Fig. 2 Carbon diffusion at 930℃ for 5 hrs.   Fig.3 Carbon diffusion at 800℃ for 24
            (a) 0T, (b) 10T.   hrs. (a) 0T, (b) 10T.

case. The diffusion temperature was set to 800℃ and the diffusion time is 24 hours. After diffusion, the sample was quenched by helium gas. At 800℃, Fe-0.8C is in austenite state, pure Fe is in ferrite state. Carbon atoms diffuse in austenite and ferrite. With the diffusion of carbon from Fe-0.8C to pure Fe, the interface between austenite and ferrite moved from the original interface of diffusion couple (white line) to right (pure Fe side). The carbon concentration at the interface jumps according to phase diagram. At the austenite side, carbon concentration is much higher than that at the ferrite side. Figure3 clearly shows that the moving distance is shorter with magnetic field of 10T than that without magnetic field. The moving speed depends on carbon diffusion in both austenite and ferrite. So the carbon diffusivity may become low in austenite and/or ferrite. Another possible reason is that magnetic field changed the phase equilibrium. According to phase diagram calculation, the A_3 line in Fe-C phase diagram moved to high carbon side with magnetic field. Therefore, equilibrium carbon content is higher with magnetic field than that without magnetic field and the concentration difference between 0.8C and equilibrium carbon content becomes smaller. Now analysis is undergoing to understand why the moving distance of interface becomes shorter in a magnetic field. Another interesting finding in Fig.3 is that the morphology of interface is different with or without magnetic field. Without magnetic field, interface is flatter than that with magnetic field. This indicates that heat treatment in magnetic field may be a new method to control interface morphology.

References
[1] H. Ohtsuka, Ya Xu and H. Wada, "Alignment of Ferrite Grains during Austenite to Ferrite Transformation in a High Magnetic Field", Materials Transactions, JIM, 41(2000), 907-910.
[2] X.J.Hao, H. Ohtsuka, P. DE Rango and H. Wada, "Quantitative Characterization of the Structural Alignment in Fe-0.4C Alloy Transformed in High Magnetic Field", Materials Transactions, 44(2003), 211-213.
[3] X.J.Hao, H. Ohtsuka, P. DE Rango and H. Wada, "Structural Elongation and Alignment in an Fe-0.4C Alloy by Isothermal Ferrite Transformation in High Magnetic Fields", Materials Transactions, 44(2003), 2532-2536.
[4] M.Shimotomai and K. Maruta, "Aligned Two-Phase Structures in Fe-C Alloys", Scripta mater. 42(2000), 499-503.
[5] H. Ohtsuka, H. Xinjiang and H. Wada, "Effects of Magnetic Field and Prior Austenite Grain Size on the Structure Formed by Reverse Transformation from Lath Martensite to Austenite in an Fe-0.4C Alloy", Materials Transactions, 44(2003), 2529-2531.
[6] J-K. Choi, H.Ohtsuka, Ya Xu and W-Y. Choo, "Effects of A Strong Magnetic Field on the Phase Stability of Plain Carbon Steels", Scripta mater., 43(2000), 221-226.
[7] X-J.Hao and H.Ohtsuka, "Effect of High Magnetic Field on Phase Transformation Temperature in Fe-C Alloys", Materials Transactions, 45(2004), p.2622-2625.
[8] S.Nakamichi, S.Tsurekawa, Y.Morizono, T.Watanabe, M.Nishida and A.Chiba,"Diffusion of Carbon and Titanium in γ-Iron in a Magnetic Field and a Magnetic Field Gradient", J. of Materials Science, 40(2005), 3191-3198.

Materials Processing under the Influence of External Fields
Edited by Qingyou Han, Gerard Ludtka, Qijie Zhai
TMS (The Minerals, Metals & Materials Society), 2007

IMPACT OF MAGNETIC FIELD ON RECRYSTALLIZATION AND GRAIN GROWTH IN NON-FERROMAGNETIC METALS

Dmitri A. Molodov

Institute of Physical Metallurgy and Metal Physics, RWTH Aachen University,
Kopernikusstr. 14; Aachen, D-52056 Germany

Keywords: Grain growth, Recrystallization, Magnetic annealing, Magnetic anisotropy

Abstract

The latest research on magnetically affected texture and grain structure evolution during recrystallization and grain growth in non-magnetic metals is reviewed. The microstructure development during grain growth in magnetically anisotropic zinc, zirconium and titanium can be effectively influenced by means of magnetic annealing. This is due to an additional driving force for grain boundary motion which arises from a difference in magnetic free energy density between differently oriented grains. Therefore each grain of a polycrystal, exposed to a magnetic field, is inclined to grow or to shrink by a magnetic force depending on the orientation of the respective grain and its surrounding neighbors with regard to the field direction. A theoretical analysis reveals that the grain growth kinetics in anisotropic non-magnetic materials is substantially modified in the presence of a magnetic field. However, the magnetic effect is not only confined to magnetically anisotropic metals, but can also be observed in a material with magnetically isotropic properties. It is experimentally shown that the application of a magnetic field substantially enhances recrystallization in cold rolled commercial aluminum alloy.

Introduction

Recrystallization and grain growth determine and control grain microstructure formation during annealing of cold worked materials which, in turn, determines their physical and mechanical properties. The fundamental process of recrystallization and grain growth is the motion of grain boundaries. Therefore an understanding of the response of grain boundaries to exerted forces is the basic requirement for a control of microstructure and properties, indispensable for the design of advanced materials for various engineering applications.

The grain microstructure evolution can be influenced by the application of an external energetic field [1,2]. In particular, grain boundary motion can be affected by a magnetic field, owing to a driving force that arises in the field due to a crystal magnetic anisotropy. In literature most experimental observations of magnetically induced changes in recrystallization and texture development relate to ferromagnetic materials [1]. However, magnetic field effects on microstructure evolution are not restricted to ferromagnetics only. As shown in the past by Mullins [3,4] and confirmed in our more recent bicrystalline experiments [5-8], also grain boundaries in non-magnetic materials (i.e. paramagnetic and diamagnetic) can be forced to move by a magnetic driving force, induced by a crystallographic anisotropy in magnetic susceptibility. Since this magnetic force does not depend on boundary properties and can be determined accurately, measurements of magnetically driven grain boundary motion gain access to the absolute value of grain boundary mobility and its dependence on the grain boundary character [5-8]. Observations of magnetically induced selective grain growth in locally deformed zinc single crystals provided unambiguous evidence that the microstructure development in magnetically anisotropic non-magnetic materials can be controlled by magnetic annealing [9,10]. Further experiments on polycrystalline cold rolled zinc [11], titanium [12,13] and zirconium [14] revealed that magnetic annealing can cause significant changes in the crystallographic texture of these materials. However, the magnetic effect is not only confined to magnetically anisotropic metals, but can also be observed in a material with magnetically isotropic properties. In particular, the application of a magnetic field substantially enhances recrystallization in cold rolled commercial aluminum alloy [15,16].

The aim of the current paper is to report on recent results of our studies on magnetically affected recrystallization and grain growth and to demonstrate that a magnetic field can be effectively utilized as an additional degree of control of texture and microstructure development in crystalline non-magnetic solids.

Magnetic driving force for grain growth in magnetically anisotropic materials

The magnetic driving force on the boundary between two adjacent uniaxial crystals with different susceptibilities along the field direction and consequently different magnetic energy densities, is given by [3]

$$p_m^b = \frac{1}{2}\mu_0 \Delta\chi H^2 \left(\cos^2\theta_1 - \cos^2\theta_2\right) \tag{1}$$

where μ_0 is the magnetic constant, $\Delta\chi = |\chi_\parallel| - |\chi_\perp|$ is the difference in susceptibility parallel $\chi_\parallel$ and perpendicular $\chi_\perp$ to the principal (or c) axis of the crystal, H is the magnetic field strength, θ_1 and θ_2 are the angles between the direction of the magnetic field and the principal axes of both neighboring grains. The sign of p_m^b depends on the magnetic anisotropy of the material ($\Delta\chi$) and the asymmetry of the spatial orientation of both neighboring grains with respect to the magnetic field direction. For a zero magnetic driving force the grain orientations do not need to be identical but merely to satisfy $\theta_1 = \theta_2$.

In a polycrystal exposed to a magnetic field each grain experiences an effective magnetic driving force that can be expressed as the difference between the magnetic free energy density of this grain ω and an average magnetic free energy density $\bar{\omega}$ of its neighboring grains $p_m = \omega - \bar{\omega}$. According to Eq. (1) this can be expressed as

$$p_m = \frac{1}{2}\mu_0 \Delta\chi H^2 \left(\cos^2\theta - \frac{\sum_n \cos^2\theta_n}{n}\right), \tag{2}$$

where θ and θ_n are the angles between the field direction and principal axes of the considered grain and its n neighboring grains.

A fundamental property of a magnetic field to influence the orientation distribution in a polycrystal, i.e. to change an existing texture or produces a preferred orientation in a structure with initially randomly oriented grains, is obviously owing to the orientation dependency of the magnetic driving force. Assuming $\Delta\chi > 0$, as this is the case for Ti, for a grain surrounded by its neighboring grains such that $\cos^2\theta - 1/n \cdot \sum_n \cos^2\theta_n < 0$, the magnetic energy density will be lower than the average energy density of the adjacent grains ($\omega < \bar{\omega}$) and in consequence the magnetic driving force ($p_m < 0$) tends to expand this grain. With or against the curvature driving force the magnetic one additionally impels each grain of a polycrystal to grow or to shrink, depending on its orientation and the orientation of its nearest neighbors with respect to the field direction. Therefore the application of a magnetic field during annealing of magnetically anisotropic titanium can bias grain microstructure evolution with respect to growth kinetics and orientation distribution of grains.

Magnetically affected texture and grain structure evolution in anisotropic metals

Texture evoluition during grain growth in a magnetic field

As mentioned above, a magnetic driving force superimposed to a curvature driving force during grain growth can bias the microstructure evolution with respect to size and orientation distribution of grains. Experimentally it has been demonstrated first on grain growth in zinc [11]. Annealing of cold rolled zinc sheet without field results in a regular texture with two symmetrical peaks in the {0002} pole figure (Fig. 1a). If, however, the annealing occurs in a magnetic field and the sample is specifically oriented with respect to the field direction, one or another texture component can be amplified that eventually results in a single-component texture (Fig. 1b,c).

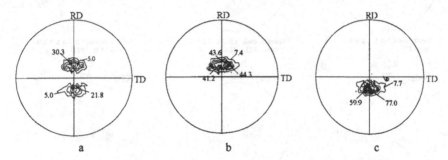

a   b   c

Figure 1. {0002} pole figures of 99%-rolled zinc-1.1% aluminum sheet samples (a) after annealing at 390°C without a magnetic field, in a magnetic field of 32 T with rolling direction (RD) tilted by +19° to the field around transverse direction (TD) (b) and tilted by -19° to the field around TD (c) [11].

Similar results, i.e. strengthening of one and weakening of another component in a sharp two-component texture, were observed in further experiments on cold rolled α-titanium [12,13]. In more details a magnetically affected texture and microstructure evolution during grain growth in commer-cialy pure titanium was investigated in our recent experiments [17].
In order to affect the texture and microstructure evolution by a magnetic field, the specimens were oriented during magnetic annealing in such a manner that the rolling direction was oriented perpen-dicular to the field, whereas the transverse direction was tilted around the rolling direction by 32° with respect to the field direction. A tilt angle of 32° ensures that one of both texture components in the (0002) pole figure is aligned perpendicular to the field direction
The cold-rolling texture with two symmetrical peaks in the (0002) pole figure (Fig. 2a) measured after 75% reduction is very similar to those already reported in literature for low alloyed titanium alloys [18]. This texture is characterized by two major ODF (orientation distribution function) peaks, widely spread around the orientations {$\varphi_1=0°$, $\Phi=35°$, $\varphi_2=0°$} and {$\varphi_1=180°$, $\Phi=35°$, $\varphi_2=0°$} [17].
The grain growth texture after 15 min and subsequent 60 min annealing at 750° is characterized by the highest ODF density at {$\varphi_1=0°$, $\Phi=35°$, $\varphi_2=30°$} and {$\varphi_1=180°$, $\Phi=35°$, $\varphi_2=30°$} orientations, although the ODF density remains high in the entire angular range of φ_2. With the proceeding grain growth at 750°C (120 and 180 min) both texture peaks in the (0002) pole figure remain nearly sym-metrical [17].
An application of a magnetic field of 17 T after 15 min conventional annealing drastically affects the texture evolution during grain growth. As also observed in our previous experiments [12], the texture components in the (0002) pole figures do not remain symmetrical anymore (Fig. 2c-f). After magnetic annealing during 60 min the ODF intensity at {$\varphi_1=180°$, $\Phi=35°$, $\varphi_2=30°$} exceeds the

correspondent intensity after annealing at zero field by a factor of 1.34, whereas the intensity at $\{\varphi_1=0°, \Phi=35°, \varphi_2=30°\}$ decreases to about 0.65 of the zero field intensity. The texture asymmetry increases with annealing time in the entire investigated time interval. After magnetic annealing during 240 min the intensity ratio of both major ODF peaks amounts to a factor of 3.5 [17].

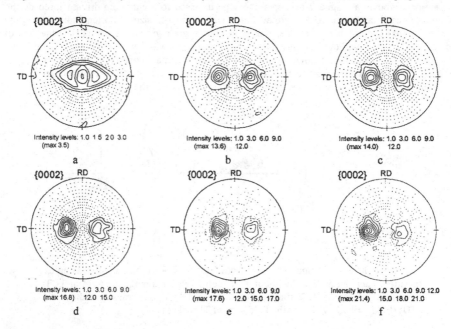

Figure 2. $\{0002\}$ pole figures for specimens (a) after 75% reduction by rolling and annealed at 750°C (b) for 75 min without field, (c) for 15 min at zero field and 60 min, (d) 120 min, (e) 180 min, (f) 240 min in a magnetic field of 17 T.

Since a magnetic field affects the intensity of all orientations with $\varphi_1=0°/180°$ and $\Phi=35°$ in the entire interval of φ_2, a most appropriate way to demonstrate the magnetically induced texture anisotropy and its increase with annealing time is to compare the $\{0002\}$ pole density of both major peaks comprising $\{\varphi_1=0°, \Phi=35°, \forall\varphi_2\}$ and $\{\varphi_1=180°, \Phi=35°, \forall\varphi_2\}$ orientations (Fig. 3).
A development of the asymmetrical texture during magnetic annealing is obviously a result of the occurrence of a magnetic driving force for grain growth. The specimens during magnetic annealing were oriented in such a manner that the c-axes of grains which comprise a texture peak around the orientations $\{\varphi_1=180°, \Phi=35°, \forall\varphi_2\}$ were directed perpendicular to the field direction. With the difference in magnetic susceptibility parallel and perpendicular to the c-axis $\Delta\chi>0$ for titanium, the magnetic free energy density of these grains approaches a minimum that results in an additional driving force for their growth ($p_m<0$).

Grain microstructure development during grain growth

A magnitude of texture peaks obtained by X-ray diffraction method is determined by the total area of grains with a respective orientation. Macrotexture measurements therefore do not provide information about the microstructure with respect to grain size and number of grains making up different texture components. However, such an analysis can be performed by utilizing the grain

22

structure reconstructed from individual orientation measurements (EBSD mapping by utilizing the Channel 5 post processing software). Since the observed texture in titanium is characterized by the high ODF density in the $\{\varphi_1=0°/180°, \Phi=35°, \forall\varphi_2\}$ orientation tubes, it is reasonable to divide the entire set of reconstructed grains into two subsets of grains with orientations around $\{\varphi_1=0°, \Phi=35°, \forall\varphi_2\}$ and $\{\varphi_1=180°, \Phi=35°, \forall\varphi_2\}$. These subsets correspond to either one of both texture peaks in the $\{0002\}$ pole figures in Fig. 2. The outcome of this approch that provides the opportunity to analyze characteristics of grains assembling the particular texture peaks is given in Table 1. Since the observed texture is rather tight a threshold grain misorientation of 3° has been used in the current analysis for the grain boundary definition to determine the mean grain sizes.

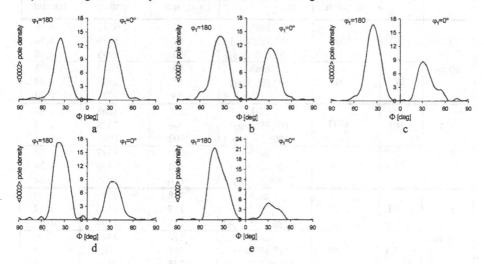

Figure 3. <0002> pole density along TD (Fig. 2) for 75%-rolled Ti sheet samples after annealing at 750°C (a) for 75 min without field; (b) for 15 min at zero field and 60 min, (c) 120 min, (d) 180 min, e) 240 min in a magnetic field of 17 T.

As seen in Fig. 3b-e, an increase of the $\{180°, 35°, \forall\varphi_2\}$ peak intensity during magnetic annealing corresponds to a decrease of the intensity of the $\{0°, 35°, \forall\varphi_2\}$ peak. Therefore it might be suggested that on average grains with the energetically favorable orientations $\{180°, 30°, \forall\varphi_2\}$ ($p_m<0$) grow at the expense of grains with the $\{0°, 35°, \forall\varphi_2\}$ orientations, those free energy is increased in a magnetic field on the amount of $p_m>0$ (Eq. (2)). Accordingly, grains with the $\{180°, 35°, \forall\varphi_2\}$ orientation can be expected to be significantly larger than grains comprising the other peak.

The results reveal that the growth kinetics, as given by the plot of the mean grain size of all grains in the reconstructed microstructure versus annealing time (Fig. 4), is enhanced by a magnetic field. In the entire investigated annealing time interval the mean grain size after magnetic annealing was found to be slightly but distinctly larger than after a conventional annealing. In both cases the growth kinetics follows the relation $d \cong Kt^n$, where d and t denote a mean grain size and an annealing time, respectively. The magnitude of the exponent n for magnetically affected kinetics $n_{17T}=0.39$, as obtained by a linear fit of the diagram in Fig. 4, was found to be slightly higher than the exponent $n_{0T}=0.37$ for non-magnetic annealings. As seen in Table 1, there is no difference in the growth kinetics for grains in different grain subsets in specimens annealed without field. In contrast, as was expected from results of macrotexture measurements, in specimens annealed in the field the mean size of grains in the subset $\{180,35,\forall\varphi_2\}$ is larger than in the subset $\{0,35,\forall\varphi_2\}$.

Figure 4b depicts that the growth kinetics in the grain subset $\{180,35,\forall\varphi_2\}$ is increased ($n_{17T}^{\{180,35,\,\text{all}\varphi_2\}}=0.44$) compared to the average kinetics over total number of grains, while this is decreased for grains in the subset $\{0,35,\forall\varphi_2\}$ ($n_{17T}^{\{0,35,\,\text{all}\varphi_2\}}=0.32$).

Table 1. Mean grain size, number of grains and fraction of different grain subsets as obtained by orientation imaging microscopy (EBSD in a SEM).

Time	Field	Total number of grains	Subset	Mean grain size, μm	Number of grains	Grain fraction
(15) + 60 min	0 T	6532	both	83	6371	0,98
			$\{0°,35°,\forall\varphi_2\}$	82	3276	0,50
			$\{180°,35°,\forall\varphi_2\}$	83	3095	0,48
	17 T	5919	both	87	5619	0,98
			$\{0°,35°,\forall\varphi_2\}$	85	2598	0,44
			$\{180°,35°,\forall\varphi_2\}$	88	3021	0,54
(15) + 120 min	0 T	4407	both	103	4268	0,98
			$\{0°,35°,\forall\varphi_2\}$	104	2328	0,54
			$\{180°,35°,\forall\varphi_2\}$	102	1940	0,43
	17 T	3539	both	109	3337	0,98
			$\{0°,35°,\forall\varphi_2\}$	102	1369	0,34
			$\{180°,35°,\forall\varphi_2\}$	116	1968	0,64
(15) + 180 min	0 T	3026	both	118	2898	0,98
			$\{0°,35°,\forall\varphi_2\}$	117	1704	0,57
			$\{180°,35°,\forall\varphi_2\}$	118	1194	0,41
	17 T	2569	both	126	2447	0,99
			$\{0°,35°,\forall\varphi_2\}$	116	992	0,33
			$\{180°,35°,\forall\varphi_2\}$	135	1455	0,66
(15) + 240 min	17 T	1976	both	139	1850	0,98
			$\{0°,35°,\forall\varphi_2\}$	126	593	0,24
			$\{180°,35°,\forall\varphi_2\}$	152	1257	0,74

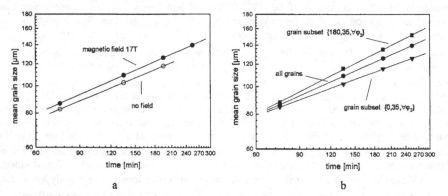

a   b

Figure 4. Mean grain size vs. annealing time (a) for specimens annealed at 750°C at zero field and in the field of 17 T and (b) in different grain subsets in specimens during magnetic annealing.

However, the observed texture anisotropy after magnetic annealings can not be completely justified by this relatively slight difference in the mean size of grains in different grain subsets. For instance, the ratio between mean areas of grains in both subsets after 240 min magnetic annealing can be estimated as $d^2_{\{180,35,\,all\varphi_2\}}/d^2_{\{0,35,\,all\varphi_2\}} = 1.45$, whereas the ratio of the pole intensity for both texture components in Fig. 3e amounts to a factor of 4.3. As given in Table 1 and illustrated in Fig. 5, a magnetic annealing also results in a drastical change of the number of grains in both grain subsets. Due to some structural inhomogeneity, probably induced during the rolling, both major texture peaks in the {0002} pole figures after annealings without field are not completely symmetric that is reflected in the non-equal number of grains in the respective grains subsets: the grain fraction in subset {0,35,$\forall\varphi_2$} is slightly bigger than this in subset {180,35,$\forall\varphi_2$} (Table 1). However, even in this case of an unequal initial state, after magnetic annealings the grains of subset {180,35,$\forall\varphi_2$} substantially outnumber the grains of subset {0,35,$\forall\varphi_2$} (Fig. 5).

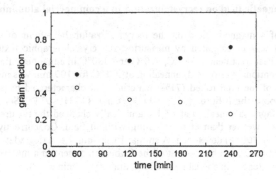

Figure 5. Grain fractions in {180,35,$\forall\varphi_2$} (solid symbols) and {0,35,$\forall\varphi_2$} (open symbols) subsets after anealings at 750°C in a magnetic field of 17 T.

Therefore, an analysis of individual orientation measurements reveals that a magnetic field affects the grain growth kinetics in such a way that the growth of grains with energetically advantaged orientations is enhanced, while the growth of grains with disadvantaged orientations is retarded. Moreover, grains with energetically advantaged orientations not only grow faster, but their fraction is larger than a half of the total grains number and continuously rises during magnetic annealing, while the fraction of grains with disadvantaged orientations decreases. This result easily explains the magnetic enhancement of growth kinetics averaged over all grains in the reconstructed grain set: the faster growth rate in a bigger and constantly rising grain fraction consequently results in increased averaged growth rate.

The observed difference in the number of grains comprising both texture peaks in the {0002} pole figures after magnetic annealing (Fig. 2c-e) can be attributed to magnetically shifted grain growth kinetics. As has been shown in Ref. [19], the Hillert expression [20] for the rate of size change of grains with size R in the presence of a magnetic field can be extended to

$$\frac{dR}{dt} = \alpha m_b \gamma \left(\frac{1}{\bar{R}} - \frac{1}{R} \right) - m_b p_m, \qquad (3)$$

where γ and m_b are the grain boundary energy and mobility (equal for all grain boundaries), and α is a dimensionless constant. In contrast to the classical Hillert approach, the mean grain size $\bar{R}$ is not the critical size of grains which neither grow nor shrink anymore. If grain growth occurs in a magnetic field, this threshold grain size R_{th} can be expressed as [19]

25

$$R_{th} = \frac{\bar{R}}{1 - \frac{\bar{R}p_m}{\alpha\gamma}} \qquad (4)$$

Due their "high energy" orientations, grains comprising the $\{0°,35°,\forall\varphi_2\}$ texture peak on average experience a positive magnetic driving force, $p_m > 0$, while for "low energy" grains of the $\{180°,35°,\forall\varphi_2\}$ peak a magnetic driving force is negative, $p_m < 0$. Therefore, according to Eq. (4), the threshold grain size for shrinkage or growth of high energy grains increases while it decreases for low energy grains. Correspondingly, more high energy grains shrink and disappear and, on the other hand, more low energy grains grow than it would be the case at zero field. Obviously this must result in an increase or decrease of fractions of differently oriented grains as observed in the experiment (Fig. 5).

Effect of magnetic field on recrystallization in a commercial aluminium alloy

Recently the effect of a magnetic field on the recrystallization behaviour of cold rolled (71%) aluminum alloy 3103 was investigated by measuring the crystallographic texture and the grain microstructure during heat treatment at 288°C, 310°C and 330°C in a magnetic field of 17 T [16]. The texture after conventional (zero field) annealing at 288°C for 100 min was measured to be very similar to the texture of the cold rolled (71%) material, and characterized by a high intensity of orientations that compose the β-fibre: {112} <111> (Cu-), {123} <634> (S-) and {011} <211> (Brass-component). Magnetic annealing at 17 T dramatically changed the texture. The deformation components were much weaker than after annealing without field. Concurrently, a high intensity Cube-component ({100} <001>) developed. A similar behaviour, i.e. a degradation of the deformation texture components and an enhancement of the Cube-component in a magnetic field were also apparent in the texture after 5 min annealing at 310°C and after 1 min at 330°C.

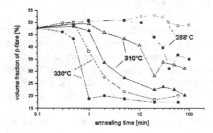

Figure 6. Volume fraction of deformation texture components calculated from the ODFs vs. annealing time at 288°C, 310°C and 330°C. Solid symbols indicate annealings in a magnetic field of 17 T, open symbols at zero field [16].

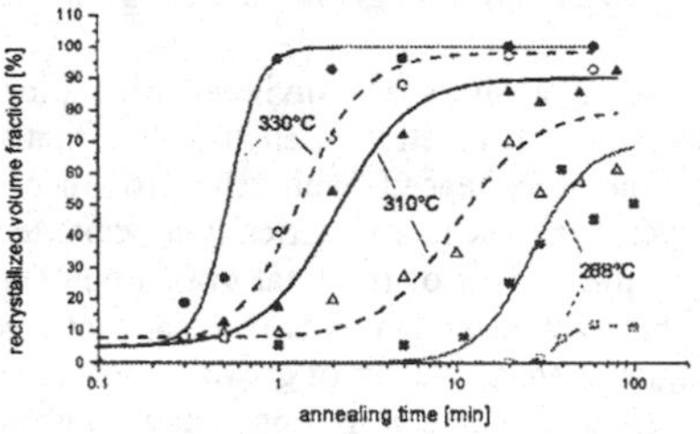

Figure 7. Recrystallized volume fraction of investigated 71% cold rolled AA3103 vs. time of annealing at zero field and in a magnetic field of 17 T. Solid symbols indicate annealings in a magnetic field of 17 T [16].

Figure 6 shows the development of the β-fibre with annealing time at different temperatures. Subsequent to an initial slight increase the volume fraction of β-fibre orientations decreased distinctly earlier and faster during annealing in a field than without field for all three investigated temperatures. A degradation of the deformation texture components is a typical sign of the progress of recrystallization in deformed Al-alloy. Correspondingly, Fig. 6 suggests that recrystallization is promoted by annealing in a magnetic field.

The change of the β-fibre volume fraction during the annealing shown in Fig. 6 can be utilized for an estimation of the recrystallization kinetics. Assuming the highest value of the β-fibre volume fraction in Fig. 6 to reflect the non-recrystallized state of the investigated material and the lowest measured value of β-fibre volume fraction to correspond to completely finished recrystallization, the recrystallized volume fraction can be calculated for all annealing times. The temporal change of the recrystallized volume fraction obtained in such a manner for all three annealing temperatures is presented in Fig. 7. Apparently, the incubation time of recrystallization during magnetic annealing is substantially decreased compared to conventional annealing. This can be attributed to magnetically enhanced recovery in the deformed aluminum alloy as observed and discussed in Ref. [15]. Moreover, the recrystallization kinetics is significantly accelerated by a magnetic field.

The reduction of the recrystallization time in a field is attributed to magnetically accelerated growth kinetics [16]. This means that the mobility of the boundaries in a deformed matrix is increased in the presence of a magnetic field. Up to now there is no any comparative study of grain boundary mobility in the presence of a magnetic field and without field. It is known, however, that a magnetic field can induce the motion of dislocations without any external mechanical stress [21,22]. Such magnetic cause of dislocation motion was attributed to spin-dependent interactions between dislocations and paramagnetic defects in the crystal structure that lead to the release of dislocations from their pinning centers [21,22]. The enhanced dislocation motion can accelerate recovery of a deformed structure as observed in Ref. [15]. The grain boundary structure can be represented by dislocation arrangements. Therefore, it is reasonable to assume that also the mobility of grain boundaries can be affected by a magnetic field.

Acknowledgements

The authors express their gratitude to the Deutsche Forschungsgemeinschaft for financial support (Grants MO 848/4-1 and MO 848/6-1).

References

1.   T. Watanabe, „External field applied grain boundary engineering for high performance materials", *Recrystallization and Grain Growth*, ed. G. Gottstein and D. A. Molodov (Springer, Berlin, 2001), 11-20.

2.   A. M. Deus, M. A. Fortes, P. J. Ferreira, and J. B. Vander Sande, „A General Approach to Grain Growth Driven by Energy Density Differences", *Acta Mater*, 50 (2002), 3317-3330.

3.   W. W. Mullins, "Magnetically Induced Grain Boundary Motion in Bismuth", *Acta Metall*, 4 (1956), 421-432.

4.   M. J. Fraser, R. E. Gold and W. W. Mullins, „Grain Boundary Mobility in Bismuth", Acta Metall. 9 (1961), 960-961.

5.   D. A. Molodov, G. Gottstein, F. Heringhaus, and L. S. Shvindlerman, „Motion of Planar Grain Boundaries in Bismuth-Bicrystals Driven by a Magnetic Field", *Scripta Mater*, 37 (1997) 1207-1213.

6.   D. A. Molodov, G. Gottstein, F. Heringhaus, and L. S. Shvindlerman, „True Absolute Grain Boundary Mobility: Motion of Specific Planar Boundaries in Bi-Bicrystals under Magnetic Driving Force", *Acta Mater*, 46 (1998) 5627-5632.

7.   A. D. Sheikh-Ali, D. A. Molodov, and H. Garmestani, „Boundary Migration in Zn Bicrystal Induced by a High Magnetic Field", *Appl Phys Letters*, 82 (2003) 3005-3007.

27

8.   P. J. Konijnenberg, D. A. Molodov, and G. Gottstein, „In Situ Measurements of Grain Boundary Migration with a High Magnetic Field Polarization Microscopy Probe", *NHMFL 2004 Annual Research Review* (NHMFL, Tallahassee, 2005), 15.
http://www.magnet.fsu.edu/publications/2004annualreport/reports/2004-NHMFL-Report137.pdf

9.   P. J. Konijnenberg, D. A. Molodov, and G. Gottstein, „Magnetically Driven Selective Grain Growth in Locally Deformed Zn Single Crystals", *Mater. Sci. Forum* 467-470 (2004), 763-770.

10.   D. A. Molodov, P. J. Konijnenberg, „Grain Boundary Dynamics and Selective Grain Growth in Non-Ferromagnetic Metals in High Magnetic Fields", *Z. Metallkd.* 96 (2005) 1158-1165.

11.   Sheikh-Ali AD, Molodov DA, and Garmestani H, „Magnetically Induced Texture Development in Zink Alloy Sheet", *Scripta Mater*, 46 (2002), 854-862.

12.   D. A. Molodov and A. D. Sheikh-Ali, „Effect of Magnetic Field on Texture Evolution in Titanium", *Acta Mater*, 52 (2004), 4377-4383.

13.   D. A. Molodov, P. J. Konijnenberg, N. Bozzolo, and A. D. Sheikh-Ali, „Magnetically Affected Texture and Grain Structure Development in Titanium" *Materials Letters*, 59 (2005), 3209-3213.

14.   D. A. Molodov, P. J. Konijnenberg, N. Bozzolo, „Effect of Magnetic Field on Grain Growth Texture in Zr", paper in preparation.

15.   D. A. Molodov, S. Bhaumik, X. Molodova, and G. Gottstein, „Annealing Behaviour of Cold Rolled Aluminum Alloy in a High Magnetic Field", *Scripta Mater*, 54 (2006), 2161-2164.

16.   D. A. Molodov, S. Bhaumik, X. Molodova, and G. Gottstein, „Magnetically Enhanced Recrystallization in an Aluminum Alloy", *Scripta Mater*, 55 (2006), 995-998.

17.   Chr. Bollmann, „Annealing Texture and Microstructure Evolution in Titanium during Grain Growth in an External Magnetic Field" (Diploma-thesis, RWTH Aachen University, 2005.

18.   A. K. Singh and R. A. Schwarzer, „Texture and Anisotropy of Mechanical Properties in Titanium and its Alloys", *Z Metallkd*, 91 (2000), 702-716.

19.   D. A. Molodov and P. J. Konijnenberg, „Grain Boundary and Grain Structure Control through Application of a High Magnetic Field", *Scripta Mater*, 54 (2006), 977-981.

20.   M. Hillert, „On the Theory of Normal and Abnormal Grain Growth", *Acta Metall*, 13 (1965), 227-238.

21.   Yu. I. Golovin, „Magnetoplastic Effects in Solids", *Physics of Solid State*, 46 (2004) 789-824.

22.   R. B. Morgunov, „ Spin Micromechanics in the Physics of Plasticity", *Physics-Uspekhi*, 47 (2004) 125-147.

28

Materials Processing under the Influence of External Fields
Edited by Qingyou Han, Gerard Ludtka, Qijie Zhai
TMS (The Minerals, Metals & Materials Society), 2007

APPLICATION OF PULSED MAGNETIC FIELD TREATMENT FOR RESIDUAL STRESS REDUCTION OF WELDED STRUCTURE

Cai Zhipeng[1], Zhao Haiyan[1], Lin Jian[1], Wang Mingdao[2]

[1]Dept. Mech. Eng.; Tsinghua Univ.; Beijing; 100084; P. R. China
[2]JingTai Tech. Co., Ltd; Wuhan; 430058; P. R. China
Keywords: magnetic treatment, stress, welded structure

Abstract

Residual stress reduction by the pulsed magnetic field treatment is attractive because the process is usually carried out at room temperature and does not lead to additional distortions. Based on the researches on the effects of magnetic field orientation and intensity, the residual stress reduction of a welded structure by pulsed magnetic field treatment was investigated. Firstly the orientation of the principal stress was investigated through the changes of magnetic susceptibility to determine the direction of the applied magnetic field. Finite element method was then used for specifying the coil current to achieve the proper magnetic intensity. Residual stress was measured by the hole-drilling method before and after magnetic treatment. Using this proposed treating process, nearly 40% residual stress was reduced and results provide first-hand information and basis for pulsed magnetic filed treatment on other similar structures.

Introduction

Pulsed magnetic treatments have been proposed as a nondestructive method to improve the mechanical properties of magnetic materials for several years. The magnetic treatments can change surface hardness [1], reduce friction [2-4], extend wear and fatigue lifetime [5-7], improve fracture and corrosion resistance [8], and retard crack growth [9]. Though the measurements of stress changes were not mentioned specially, most previous researches stated that release of residual stress during magnetic treatment was responsible for the so various improvements.

Barney E. Klamecki [10] and F. Tang et al.[11-15] studied magnetic effects on stress relief. In their experiments the residual stress was reduced by 4-13% and 40% respectively. F. Tang considered that the dynamic magnetostriction, which provides energy to set off inelastic strains via magnetoelastic coupling, was the major cause of the stress relief. They guessed that the increasing amplitude of magnetostriction would lead to more relief on stress.

In this paper whether the increasing amplitude of magnetostriction would lead to more relief on stress and how orientations affect were studied. Based on the effects of magnetic filed intensity and orientation, the residual stress reduction of a welded structure due to pulsed magnetic field treatment was investigated. Residual stress was measured by the hole-drilling method. Through the magnetic field treatment nearly 40% residual stress was reduced and this result helps to apply pulsed magnetic filed treatment on other similar structures.

1 Effects of magnetostriction and orientation in magnetic field treatment

1.1 Effects of magnetostriction on residual stress relief

In F. Tang's [15] magnetic field treatments nearly 40% residual stress was reduced. They proposed that the greater amplitude of magnetostriction would lead to more relief of stress. However, we found that with the increase of the applied magnetic field, the amplitude of the magnetostriction, with directions of both parallel and perpendicular, became accordingly greater. Stress relief did not obey the similar rules. There existed a range of magnetic field which were helpful for residual stress reduction. Beyond this range, the stress would remain unchanged, or even become increasing. Moreover, strains caused by magnetostriction were always in elastic range, which usually means stress had little chance to be altered. In this paper nearly 40% residual stress was reduced throuth a designative magnetic field intensity, which provided a reference value in other magnetic field treatment. More details were shown in reference [17].

Fig.1 and Fig.2 show magnetstriction and the corresponding stress variation respectively.

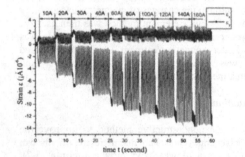

Fig.1   Magnetostriction with different coil currents

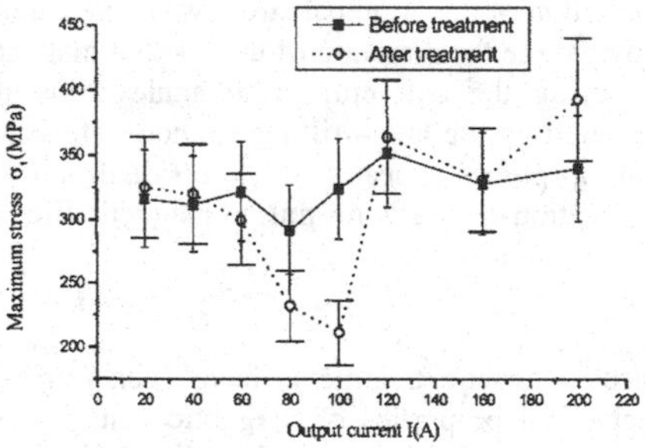

Fig.2   Relation between maximum output current and change of principal stress σ_1

Based on the measurement data, there should be a range of magnetic intensity for stress relief and stronger magnetic field does not indicate more stress reduction.

1.2 Orientation effects in pulsed magnetic treatment

The orientation effect on residual stress relief by pulsed magnetic treatment was another critical issue. In this paper specimens were produced by Tungsten Inert Gas (TIG) welding and the initial residual stresses were measured by the hole-drilling method. These specimens were treated in the same magnetic field but different orientations and the final residual stresses were measured after treatment. Fig.3 shows the orientation effects. By comparing the stress values before and after treatment, magnetic treatment effects were significantly related to orientations.

30

For case 1 the magnetic treatment had no effect on stress relief. The stress reduction was within 5%. For case 3 the reduction was the most remarkable and 26% residual stress was relieved. For case 2 the effect was in between and the reduction was about 17%.

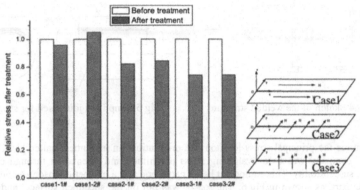

Fig.3 Stress relief with different magnetic orientations

Based on the experiments, magnetic treatments were remarkably influenced by orientations. More residual stress could be reduced through a magnetic field perpendicular to the principal stress.

2   Procedure of magnetic field treatment for a welded structure

From the results mentioned above, both proper magnetic intensity and orientation were necessary for obvious stress relief in a magnetic field treatment. As for a real welded structure, the geometric complexity made the calculation of magnetic induction by theoretical formula nearly impossible. The similar problem exited in how to designate the magnetic field orientation, because stress was usually complex and hard to be measured.

In this paper, magnetic induction was calculated through Finite Element Method (FEM). Stress directions were measured by equipment which recorded the changes of magnetic susceptibility with stress.

2.1 Introduction of the welded structure

A welded structure with a length of 3200 mm used as instrument support was shown in Fig.4. Because this structure was made of steel tubes and armor plates, and all the parts were welded together, large stresses were generated during welding process and the residual stress usually caused severe distortions. Vibrating method was usually used to reduce stress and improve geometrical stability. Sometimes cracks were observed in the weld beads during vibrating and the whole structure had to be repaired afterward. Since heat treating was unacceptable for metallurgical and economic reasons, manufacturer looked for a new method to reduce stress efficiently. Magnetic field treatment shows some advantages over other methods.

2.2 Designation of magnetic field orientation

Since residual stress was mainly located in the weld bead, a simplified joint was used studying stress relief as shown in Fig.5.

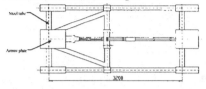

Fig.4 Sketch of the welded structure   Fig.5 Simplified joint used for stress relief study

To determine the principal stress location and orientation, an instrument called "MKP-102" by Nikkoshi company was used evaluating stress magnitude and orientation through changes of magnetic susceptibility. This equipment had 2 parts: one was a magnetic sensor, the other was an electronic part, as shown in Fig.6. The sensor was put on the specimen surface, and the index showed the relative value of stress at the measurement position in the specified direction. Usually the relative value was an average stress with a depth of 5mm. The sensor was then moved and rotated to find the principal stress on the surface.

This method was convenient and flexible, but higher precision was needed to evaluate magnetic field treatment effects. So hole-drilling method was also conducted to measure the stress. As for determining only the stress orientation, the former instrument would be accurate enough.

Fig. 6 "MKP-102" measuring residual stress

Both the magnetic method and the hole-drilling method showed that the maximum stress was located in the weld bead and the principal stress was perpendicular to the weld bead.

Based on the conclusion in Ref. [16], if applied magnetic field was perpendicular to the direction of principle stress, it had potential to release more residual stress. And the applied magnetic field should be aligned with the welding bead direction for this structure.

2.3 Choosing the magnetic intensity

Ref. [17] showed that there was a range of magnetic intensity for stress relief. Residual stress kept unchanged or even increased after magnetic field treatment beyond the range. In this paper, ANSYS program was used to calculate the coil current for the required intensity.

The numerical model was shown in Fig.7.

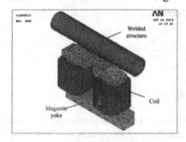

Fig.7 FEM model for magnetic intensity calculation

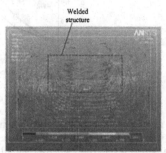

Fig.8 Vector field of magnetic intensity with the exiting current of 100A

Fig.8 showed that when the coil current reached 100 A, the magnetic density in the weld bead would be best for stress relief.

3   Results and Discussion

Fig.9 showed the measured positions by the hole-drilling method before and after magnetic field treatment. The stress changes caused by the magnetic field treatment were shown in Fig.10.

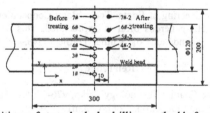

Fig.9 Surveyed positions of stress by hole-drilling method before and after treatment

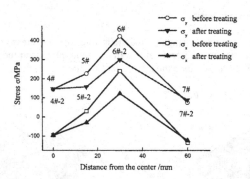

Fig. 10 Stress changes caused by magnetic field treatment

It was found that after the magnetic field treatment the principal stress was reduced by 120MPa, which was significant for this structure. It was also found that released stress was related to the original stress level. The higher the initial stress, the more stress released. If the original stress was not high enough, the method presented in this paper had no effects on the stress relief, which was much different from heat treatment method.

If this method was generally used in real products, several technical problems had to be resolved, such as how to designate the treating time, whether the treatment with a few strong magnetic pulsed equaled to the one with more but weak pulses. And whether the treated parts had an influence on the parts to be treated also needed to be further studied.

4 Acknowledgements

Wang Mingdao, a senior engineering in Jingtai Technology Cooperation, gave us lots of aids in equipment manufacturing. His generous offer is appreciated. This work is also supported by National Nature Science Foundation of China (Project No. 50505019). Their help is acknowledged.

Reference

[1] D. Paulmier: Surface and Coating Technology, 1995, v76-77, 583-588

[2] Paul C. Miller: Tooling and Production, 1990, v55, n12,100-103

[3] R. F. Hochman, N. Tselesin and V. Drits: Advance Materials and Processes, 1988(8), 36-41

[4] Omar Batainech, Barney Klamechi and Barry G. Koepke: Journal of Materials Processing Technology, 2003, v134, n2, 190-196

[5] M. S. C. Bose: Phys. Stat. Sol. (a), 1984, v86, 649-654

[6] Yusef Fahmy, Tom Hare, Robert Tooke and Hans Conrad: Scripta Materialia, 1998, v38, n9, 1355-1358

[7] Lu Baotong, Qiao Shengru and Sun Xiaoyan: Scripta Materialia, 1999, v40, n7, 767-771

[8] Snegovskij F.D and Uvarov V. A.: Trenile I Iznos n3, May-Jun. 1991,535-539

[9] S. N. Prasad, P. N. Singh, V. Singh: Scripta Materialia, 1996, v34, n12, 1857-1860

[10] Barney E. Klamecki , Journal of Materials Processing Technology , 2003 , v141, n3, 385–394

[11] F. Tang, A.L. Lu, H.Z. Fang, J.F. Mei: Materials Science and Engineering A, 1998, v248, 98–100

[12] A. L. Lu, F. Tang, X. J. Luo, J. F. Mei, H. Z. Fang: Journal of Materials Processing Technology, 1998, v74, 259-262

[13] F. Tang, A. L. Lu, J. F. Mei, H. Z. Fang, X. J. Luo: Journal of Materials Processing Technology, 1998, v74, 255-258

[14] F. Tang, A. L. Lu, H. Z. Fang, X. J. Luo: Transaction of the China Welding Institution. 2000, v21, n2, 29-31

[15] F. Tang. Research on Residual Stress Reduction in Steel by Pulsed Magnetic Treatment. Dissertation of P.H.D. Department of Mechanical Engineering, Tsinghua University. 1999

[16] Z.P. Cai, H.Y. Zhao, J. Lin: Materials Science and Engineering A, 2005, v398, 344-348

[17] Z.P. Cai, J. Lin, H.Y. Zhao.: Materials Science and Technology, 2004, v20, 1563-1566

Materials Processing under the Influence of External Fields
Edited by Qingyou Han, Gerard Ludtka, Qijie Zhai
TMS (The Minerals, Metals & Materials Society), 2007

RETAINED AUSTENITE IN SAE 52100 STEEL POST MAGNETIC PROCESSING AND HEAT TREATMENT

Nathaniel Pappas[1], Thomas R. Watkins[2], O. Burl Cavin[3], Roger A. Jaramillo[2],
Gerard M. Ludtka[2]

[1]Ball State University, 2000 W. University Ave, Muncie, IN 47306
[2]Materials Science and Technology Division, Oak Ridge National Laboratory
One Bethel Valley Road, P.O. Box 2008, MS 6064, Oak Ridge, TN 37831
[3]Center for Materials Processing, University of Tennessee, Knoxville, TN 37996

Keywords: Retained austenite, Martensite, Magnetic Processing, X-ray Diffraction

Abstract

The application of magnetic fields as a method of reducing retained austenite in steel is based on the thermodynamic argument that paramagnetic austenite is destabilized in the presence of a large magnetic field. The goal of this study was to determine the effect of applying a magnetic field at room temperature after quenching on the amount of retained austenite. After heat treatment and several post-quench delay times, samples of SAE 52100 steel were heat treated then subjected to a magnetic field of varying strength and time. X-ray diffraction was used to collect quantitative data corresponding to the amount of each phase present post processing. Stronger magnetic field strengths resulted in lower amounts of retained austenite for fixed application times. The results for applying a fixed magnetic field strength for varying amounts of time were inconclusive and suggest negligible dependence. When applying a magnetic field after a post-quench delay time, the analyses indicate that decreased delay times result in less retained austenite, which indicates that retained austenite becomes more stable with time after quenching. Established martensite transformation kinetic equations were applied to estimate the effect of the magnetic field on martensite start temperature and the resulting amount of retained austenite. A data fit indicates a 2°C/Tesla shift in martensite transformation kinetics.

Introduction

Multiple researchers [1-3] have shown that the application of high or ultrahigh magnetic fields can alter the microstructure and phase transformations in medium and high carbon steels such as SAE 52100. Because of the high carbon content in SAE 52100, retained austenite can result from the heat treatment process. Since retained austenite can transform with a nominal 4% volume expansion into martensite, lower amounts of retained austenite are usually sought in order to avoid any corresponding distortion and/or loss of fracture toughness in the final piece. Typically, a post heat treatment tempering cycle or a cryogenic treatment immediately following quenching will minimize the amount of retained austenite, but this is time-consuming and costly [4]. The application of a magnetic field as a method of reducing the amount of retained austenite in steel may overcome these difficulties and is based on the thermodynamic argument that paramagnetic austenite is destabilized in the presence of a large magnetic field.

Samples of SAE 52100 steel, which contain 1.45 wt% Cr and 1.0 wt% C, were studied to determine the relationship between applying a magnetic field post quench and the amount of retained austenite present at room temperature [5]. This project is a part of a larger, ongoing project investigating the application of a magnetic field during heat treatment and its influence on the iron-carbon phase-equilibria.

Experimental Procedures

A 32 mm diameter bore resistive magnet with a maximum magnetic field strength of 33 Tesla (T) was used at the National High Magnetic Field Laboratory for the magnetic processing of the samples. The custom designed induction coil with a gas purge/quench system was used in the heat treatment of the samples and is described in [1, 3]. All samples were heated to 850° C and held for 30 minutes to insure full transformation into austenite. The samples were then quenched to 25° C using He gas. Magnetic fields with strengths ranging from 0 to 30 T were applied post quench at room temperature to the samples for varying times either immediately following the quench or after a wait time.

Table I lists the details of the experimental conditions for the x-ray measurements. Briefly, a 4-axis (Φ, χ, Ω, 2Θ) goniometer [6] was employed for phase identification using Θ-2Θ scans and near-parallel beam optics, which reduces sample surface displacement errors [7,8]. During scanning, the specimens were oscillated ±2 mm in plane to improve particle statistics. Specimen alignment was accomplished using a dial gauge probe, which was accurate to ±5 µm. Here, the relative distance to the center of rotation is known, and the diffracting surface is positioned accordingly. The lattice parameter of LaB_6 powder on a zero background plate was determined to be 5.156(3) Å and compared favorably with 5.157 Å [9], demonstrating reasonable goniometer alignment. Jade software [10] was utilized in the phase identification analysis. In order to get a measure of the reproducibility, one sample was scanned 5 times in different locations/sets of grains. The amount of retained austenite was then calculated, and the standard deviation determined to 1.1 v%, which was assumed to be representative for all the samples. Five Vicker's indents (500 g load) were averaged for each hardness value.

Table I - Experimental conditions of the x-ray measurements.

Parameter	Condition
Equipment	Scintag PTS goniometer
	Spellman DF3 series 4.0 kW generator
	Scintag liquid N_2-cooled Ge detector
Power	1.4 kW; 40 kV, 36 mA
Radiation	Cr, λ = 2.28970 Å
Incidence slit divergence	0.5°
Receiving slit acceptance	0.25°; radial divergence limiting (RDL) Soller slit
Source to specimen distance	290 mm
Specimen to back slit distance	290 mm
Scans	0.02° 2Θ/step at 0.1 or 0.5 °/min

The amount of retained austenite was calculated using the ASTM standard for randomly oriented samples [11] where R is a scale factor associated with phases and materials used, I is the calculated integrated intensity total for the phase present, and V is the volume fraction of the phase. The austenite phase is represented by γ and the ferrite/martensite phase is represented by α.

$$\% of\ RA = (V_\gamma)100 = \left(\frac{\frac{I_\gamma}{R_\gamma}}{\frac{I_\gamma}{R_\gamma} + \frac{I_\alpha}{R_\alpha}} \right) 100 \tag{1}$$

The samples in this study contained "white" or untempered martensite, and the body centered tetragonal nature (BCT) may be discerned in the XRD patterns (see Figure 1). The ASTM standard was written for tempered martensite because untempered martensite has almost no practical application. Since the tempering process reduces the body-centered tetragonality of the martensite, the x-ray diffraction pattern for the tempered martensite typically appears body centered cubic (i.e., BCC like ferrite). Thus for this work, the ASTM standard was modified such that the BCT peaks were fit as the sum of two peaks to account for the tetragonal distortion. That is, the BCT peak areas for the (002) + (200) and the (112) + (211) were simply substituted for the BCC (200) and (211), respectively, with no other modification to the analysis. The peak areas for FCC (200) and (220) were used per the ASTM standard to calculate the retained austenite. Further the samples were taken to have only two phases. The errors associated with these modifications were assumed to be small.

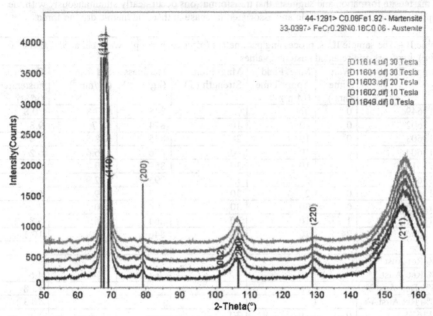

Figure 1. The XRD patterns of 52100 samples subjected to 0, 10, 20, and 30 T magnetic fields for 10 min immediately after quenching.

Results / Discussion

Table II lists the hardness and amount of retained austenite for the samples examined. As a check, several NBS standards were analyzed. The results were found to be in good agreement with the standards within the error of the measurement. The control sample, D11649, contained ~10.8 v% retained austenite. In Figures 1 and 2, the percentage of retained austenite can be seen to decrease as a function of magnetic field strength. Applying Andrews equation for chemistry dependent martensite start temperature (Ms) [12] and the Koistinen-Marburger (K-M) equation for martensite transformation kinetics [13], an estimate for the volume fraction of retained austenite can be made (see curve in Figure 2). Assuming that the volume fraction of cementite remains constant at 5% and the concentration of carbon in austenite is 0.7 wt. % prior to the quench, Ms is estimated to be 227°C and retained austenite is ~11 vol. %. K-M kinetics provide an estimate of changes in retained austenite with applied field. This effect is explicitly introduced as an increase in Ms temperature. Such a relationship is given in the equation below:

$$f^{\gamma} = e^{-0.011\left(Ms + \Delta Ms - 25°C\right)}. \qquad (2)$$

Here, f is the volume fraction of austenite, and ΔMs is the apparent increase in Ms due to the magnetic field. A linear relationship is assumed, $\Delta Ms = L H$, where H is the magnetic field in Tesla and L is the coefficient. A fit of the measured data suggests a coefficient of 2.0°C/Tesla. In Figure 3, the amount of retained austenite appears to be independent of the amount of time the magnetic field is applied. These results in conjunction with the results in Figure 2 indicate that it isn't how long the magnetic field is applied for, but the strength of field that has an effect on the amount of retained austenite present. This is consistent with the athermal transformation kinetics of martensite formation and suggests that transformations occur nearly simultaneously with the application of the magnetic field and associated increase in thermodynamic driving force.

Table II – The sample ID's, processing parameters (after each sample was held at 850°C for 30 min.), hardness and retained austenite values.						
Sample ID	Wait Time (hrs.)	Mag. Field Apply Time (Min.)	Mag. Field Strength (T)	Hardness (kg/mm^2)	Std. Dev. (kg/mm^2)	Retained austenite (v%)†
D11649	0	0	0	845	±27.8	10.8
D11602	0	10	10	874	±8.7	9.8
D11603	0	10	20	875	±28.5	7.4
D11604	0	10	30	876	±16.6	6.2
D11614	0	10	30	859	±19.7	4.7
D11612	0	2	30	895	±22.9	6.4
D11613	0	2	30	897	±19.8	6.3
D11605	0	30	30	894	±10	7.2
D11601	1	10	30	864	±16.1	5.8
D11606	24	10	30	846	±23.9	10.1
2% Ret Aust.*						3.1
5% Ret. Aust.**						4.6
15% Ret. Aust.¥						13.8
30% Ret. Aust.¥¥						30.3
D11614-1	0	10	30			3.1
D11614-2	0	10	30			5.0
D11614-3	0	10	30			2.7
D11614-4	0	10	30			3.0
D11614-5	0	10	30			2.2

† standard deviation ±1.1%; * NBS 488 2% austenite in ferrite; ** NBS 485a 5% austenite in ferrite; ¥ NBS 486 15% austenite in ferrite; ¥¥ NBS 487 30% austenite in ferrite

Figure 4 plots retained austenite as a function of wait times after quenching before the magnetic field is applied. The large value at long wait time indicates that the retained austenite is stabilizing over time and that the magnetic field is less effective at destabilizing the paramagnetic austenite. Although more data are required to determine the kinetics of retained austenite stabilization during post quench wait time, it is clear that waiting 24 hours to apply a magnetic field significantly reduces its ability to reduce the retained austenite percentage. The hardness was measured for comparison to other magnetically processed steels notreport ed here. The hardness data in Figure 5 showed no trends, indicating that the magnetic field has little effect on the hardness of the 52100 samples.

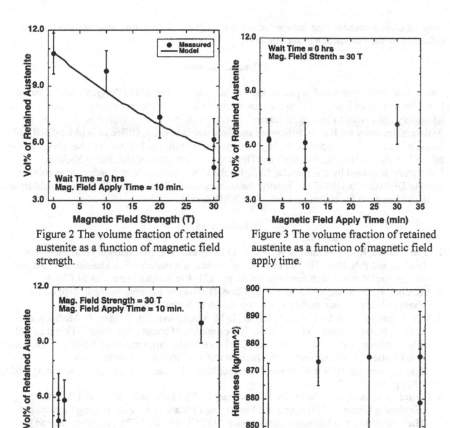

Figure 2 The volume fraction of retained austenite as a function of magnetic field strength.

Figure 3 The volume fraction of retained austenite as a function of magnetic field apply time.

Figure 4 The volume fraction of retained austenite as a function of post quench wait time.

Figure 5 The hardness as a function of magnetic field strength.

Summary & Conclusion

X-ray measurements of retained austenite in annealed and quenched SAE 52100 steel indicate that room temperature application of a magnetic field can reduce the amount of retained austenite. Further, data suggest delay of the application of the magnetic field allows the austenite to stabilize. These data indicate that what is important is not how long the magnetic field is applied, but rather the strength of the magnetic field that is applied. Due to the fact that the austenite tends to stabilize over time, if magnetic processing is to be used as a viable method for retained austenite reduction, the field must be applied soon after the quench. The strength of a magnetic field seemed to have no effect on the hardness of the SAE 52100 steel. Using the Koistinen-Marburger equation for martensite transformation kinetics, it is estimated that the magnetic field increases Ms by approximately 2°C/Tesla. This project is a part of a larger, ongoing project investigating the application of a magnetic field during heat treatment and its influence on the iron-carbon phase-equilibria.

Acknowledgements

This research was sponsored in part by the Pre-Service Teacher (PST) Program, funded by the U.S. Department of Energy to give the opportunity to aspiring teachers to work at a national laboratory and conduct meaningful research. This research was also sponsored in part by the Assistant Secretary for Energy Efficiency and Renewable Energy, Office of FreedomCAR and Vehicle Technologies, as part of the High Temperature Materials Laboratory User Program and by the Laboratory Directed Research and Development Program of Oak Ridge National Laboratory, managed by UT-Battelle, LLC, for the U.S. Department of Energy under contract number DE-AC05-00OR22725. Finally, the authors would like to thank C. Correa-Lockhart and B. B. Witherspoon for collecting most of the XRD data.

References

1. G.M. Ludtka, R.A. Jaramillo, R.A. Kisner, D.M. Nicholson, J.B. Wilgen, G. Mackiewicz-Ludkta, and P.N. Kalu, "In situ evidence of enhanced transformation kinetics in a medium carbon steel due to a high magnetic field," pp. 171-4 in Scripta Materialia 51 (2004).
2. J-K Choi, H. Ohtsuka, Y. Xu, and W-Y. Choo, "Effects of a strong magnetic field on the phase stability of plain carbon steels," pp. 221-6 in Scripta Materialia 43 (2000).
3. R.A. Jaramillo, G.M. Ludtka, R.A. Kisner, D.M. Micholson, J.B. Wilgen, G. Mackiewicz-Ludtka, N. Bembridge, and P.N. Kalu, "Investigation of phase transformation kinetics and microstructural evolution in 1045 and 52100 steel under large magnetic fields", in Solid-to-Solid Phase Transformations in Inorganic Materials 2005, Vol. 1: Diffusional Transformations, (Eds. J.M. Howe, D.E. Laughlin, J.K. Lee, U. Dahmen, W.A. Soffa), TMS, PA, pp. 893-898 (2005).
4. Gerard M. Ludtka, R.A. Jaramillo, R.A. Kisner, G. Mackiewicz-Ludtka and J.B. Wilgen, "Exploring Ultrahigh Magnetic Field Processing of Materials for developing Customized Microstructures and Enhanced Performance," ORNL/TM-2005/79, [Technical Report] available at website: http://www.osti.gov/bridge.
5. Harry Chandler, *Metallurgy for the Non-Metallurgist*, Materials Park, Ohio: ASM International, 1998, p.162.
6. H. Krause and A. Haase, "X-Ray Diffraction System PTS for Powder, Texture and Stress Analysis," pp. 405-8 in Experimental Techniques of Texture Analysis, DGM Informationsgesellschaft•Verlag, 1986, H. J. Bunge, Editor.
7. I. C. Noyan and J. B. Cohen, Residual Stress, Measurement by Diffraction and Interpretation, Springer-Verlag, New York, 1987, pp. 204-5.
8. T. R. Watkins, O. B. Cavin, J. Bai and J. A. Chediak, "Residual Stress Determinations Using Parallel Beam Optics," pp. 119-129 in Advances in X-Ray Analysis, V. 46 CDROM. Edited by T. C. Huang *et al.*, ICDD, Newtown Square, PA, 2003.
9. Card # 34-427, PDF file, International Centre for Diffraction Data, Newtown Square, PA.
10. Jade 6.5-XRD Pattern Processing, MDI, Livermore, CA 94550, USA.
11. Standard practice for X-ray determination of retained austenite in steel with near random crystallographic orientation, American Society for Testing and Materials, Designation: E 975-95 (1999) 1-6.
12. K. W. Andrews, " Empirical Formulae for the Calculation of Some Transformation Temperatures," *J. Iron Steel Inst.*, 203, 721-727 (1965).
13. D. P. Koistinen and R. E. Marburger, "A General Equation Prescribing the Extent of the Austenite-Martensite Transformation in Pure Iron-Carbon Alloys and Plain Carbon Steels", *Acta Met.*, Vol. 7, 1959, pp. 59-60.

Materials Processing Under the Influence of External Fields

Session II

METALS SOLIDIFICATION UNDER THE INFLUENCE OF EXTERNAL FIELDS: RESEARCH STATUS IN SHANGHAI UNIVERSITY

Qi-Jie Zhai, Yu-Lai Gao

Center for Advanced Solidification Technology (CAST), Shanghai University; Yanchang Road 149, ZhaBei District; Shanghai, 200072, P.R. China

Keywords: Metals solidification, External fields, Application

Abstract

In the present study the research status of metals solidification under the influence of external fields in Shanghai University, which is one of the main academic institutions in China, was introduced. The special self-developed equipments to produce external fields were presented. At present, many kind of external fields have been developed in Shanghai University, including ultrastrong magnetostatic field, ultrasonic, electric current pulse, pulsed magnetic field, and magneto oscillation etc. Also, the research interests in the metals solidification under external fields were systematically illuminated. Besides, the application prospect of these techniques was reviewed, especially in the metallurgy industry.

Introduction

One of the important aspects of materials science is the control of solidification process of liquid metals since the solidified microstructure has a profound influence on the mechanical properties. With the development of modern science and technologies the requirement for high-performance and environment friendly materials has become increasingly urgent. Environmental harmony of materials promotes the generation of green metallurgy. Conventional grain refinement processes such as modification and alloying treatment undergo great challenges. The solidification process develops in the direction of simplification, saving, high efficiency and environmental protection. The control of solidification process by several types of physical fields satisfies such requirement and facilitates the development of new solidification techniques. Of special significance in the field of metal solidification is the widespread application of electromagnetic fields with the aim of obtaining high-quality products with particular solidified microstructures.

Electromagnetic processing of materials (EPM) refers to an application of electronic or magnetic field or both to the solidification of liquid metals, controlling the solidification behavior and final microstructures and mechanical properties. In the last several decades, much attention has been paid to the effect of physical fields on the solidification process and solidified microstructures, as exemplified for the effect of electrical current on the solidified microstructure[1], electromagnetic processing of materials [2], ultrasonic grain refinement [3, 4] etc. With the features of pollution free of the external fields and characteristics of materials, the microstructure can be promoted and the resultant mechanical properties is markedly increased. Therefore, EPM is a promising materials processing technique with great potentials.

From 1990s, several research groups in Shanghai University began to conduct the study of materials processing under the external fields, and they have employed many types of external fields, including high frequency amplitude-modulated magnetic field, ultrastrong magnetostatic field, ultrasonic, electric current pulse, pulsed magnetic field and magneto oscillation etc. Depending on these external field techniques, extensive investigations have been carried out not only on fundamental researches but also on application studies. In this paper, the research status of metals solidification under the influence of external fields should be briefly reviewed.

Research status

The main investigations and progress made by our research group and other researchers in Shanghai University are introduced as follows:

Soft-contact electromagnetic continuous casting

Chosen pure Sn, the influence of high frequency magnetic fields on surface quality of billets in the soft-contacted electromagnetic continuous casting was investigated by the equipment shown in Figure 1, and several kinds of high frequency amplitude-modulated magnetic fields including rectangle, triangle and sine wave ones were adopted. The results showed that when the modulated wave frequency was a little lower than the intrinsic frequency of the experimental system, the intermittent contacting distance was the greatest, the mould flux lubrication was the best, the continuous casting with drawing resistance was the least and the surface quality of billets was the best. Furthermore, sine wave was the best in deducing the withdrawing resistance and improving the billets surface quality among the aforementioned three kinds of amplitude-modulated magnetic fields [5-7].

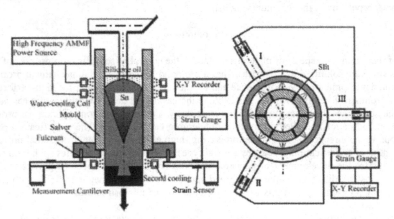

Figure 1. Experiment apparatus of mould oscillation-less continuous casting under high frequency amplitude-modulated magnetic field [5].

Metals solidification under the influence of ultrastrong magnetostatic field

Fundamental studies of the ultrastrong magnetostatic field on metals solidification process have been conducted on the experimental device (shown in Figure 2) in many alloy systems, such as Al-18%Si, Cu-Pb, Pb-Sn, Al-Ni, Zn-Cu etc. The research interests include the influence of magnetostatic field on the morphology of primary crystal, solidification structure, and growth orientation of directional structure etc. The magnetic intensity ranges from $0 \sim 14T$, which can be continuously adjusted.

In particular, the influence of ultrastrong magnetostatic field on the solidification structure of some special compound such as Mn-Bi was carried out. The results showed that the BiMn grains accumulated on the periphery of the treated specimen, forming a ring-like MnBi phase-rich layer where MnBi phase aligned along the magnetic direction [8, 9]. Furthermore, great effects can impose to the growth of alloys. According to the Bridgman growth of Al-4.5wt%Cu alloys, the field causes disorder dendrites and their tilt in orientation. The ultrastrong field tended to alter growth direction of the dendrites from <100> to <111> at lower grow speed, and the extent of the alternation depended on the intensity of magnetic field and the growth speed of the crystal [10].

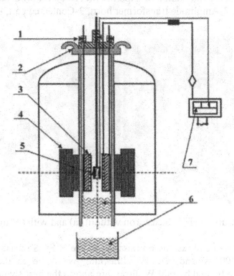

Figure 2. Schematic diagram of experimental device of metal solidification under the ultrastrong magnetostatic field.
1-Sample frame, 2-Water-cooling cover, 3-Heat furnace, 4-Superconductor magnet, 5-Sample, 6-Water barrel for quenching, 7-Temperature control system

Metals solidification under the influence of ultrasonic[11-14]

The output power of the ultrasonic generator is 0 ~ 1000W, and the frequency ranges from 17 ~ 23kHz. In order to avoid the ultrasonic erosion of the amplitude transformer horn and the pollution to the melt, a non-contact technique has been developed (shown in Figure 3).

Applying the said setup, the influence of ultrasonic on solidification process has been conducted in the alloy systems include Sn-Sb, ZL101, HT150 cast iron, 1Cr18Ni9Ti stainless steel, Q235 steel and T10 steel etc. It was found that the solidification structure was obviously refined under the effect of ultrasonic. The higher ultrasonic power, the finer the microstructure. Also, the gravity segregation in the Sn-Sb alloy was greatly improved, and homogeneous, refined structure was obtained. In particular, the graphite changed from the flake-shape to granular one after the ultrasonic treatment (shown in Figure 4).

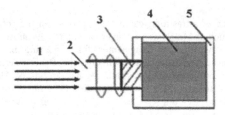

Figure 3. Schematic illustration of experimental setup for ultrasonic treatment.
1- Ultrasonic vibration, 2-Amplitude transformer horn, 3-Contacted part, 4-Melt, 5-Mould.

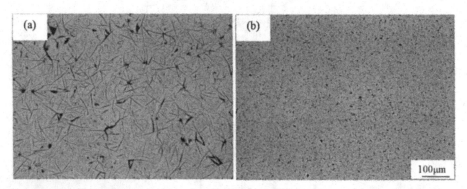

Figure 4. Microstructure of HT150 cast iron without (a) and with (b) ultrasonic treatment.

Figure 5 shows the effects of ultrasonic on microstructure of Sn-Sb peritectic alloy. Comparing the effects of 0W, 400 W and 600 W ultrasonic, we come to the conclusion that the microstructure of sample treated by 600 W ultrasonic shows the best shape and size. Besides the structure refinement, the gravity segregation has been greatly improved.

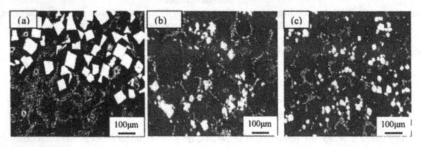

Figure 5. Microstructure of Sn-Sb alloy and the influence of ultrasonic: (a) 0W, (b) 400W, (c) 600W.

Figure 6 shows the results of the structure under various treatment. It is obviously seen that the grain sizes decrease from 8560μm without any treatment to 386μm treated by the ultrasonic with the power 600W. Moreover, better refinement was obtained if the specimen was treated by the ultrasonic than that treated by the modificator.

For T10 steel, even homogeneous equi-axed grains were gained under the effect of ultrasonic with different power. And it seems that high ultrasonic power is favorable to refine the solidification structure of T10 steel (shown in Figure 7). Attributing to the influence of the ultrasonic, the grain size was decreased from 3210μm to 323μm when the ultrasonic power arrive to 600W.

With respect to the intrinsic attenuation of the ultrasonic during its propagation in the melt, the attenuation regularity was investigated in Sn-Sb alloy. The relation of the ultrasonic attenuation versus the propagation distance and melt temperature was approximately expressed by,

$$I = \exp(4.7 - 0.02x) - 0.15\exp\left\{-1.7(x-27)^2\right\} \cdot T^{(0.001x+0.63)} \qquad (1)$$

where I is the ultrasonic intensity, x the propagation distance and T the melt temperature.

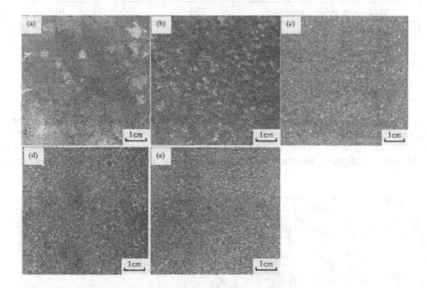

Figure 6. Solidification structure of ZL101 Al-Si alloy without (a) and with the treatment of modificator (b) and ultrasonic with the power of 200W (c), 400W (d) and 600W (e).

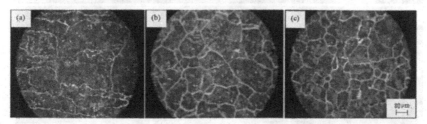

Figure 7. Solidification structure of T10 steel under the influence of ultrasonic with the power of (a)0W, (b)400W and (c) 600W.

Metals solidification under the influence of electric current pulse [15-17]

Three power sources have been designed in our research group. These power sources are divided into medium-voltage and high-capacity pulse power source (PSI), high-voltage (PSII), and medium-frequency (PSIII) ones. The parameters of the self-developed power sources are elaborately listed in Table I.

Table I Parameters of the power sources PSI, PSII and PSIII.

Parameter	PSI	PSII	PSIII
Maximum electric capacity/μF	7500	2000	1500
Highest nominal voltage/kV	3	10	500
Maximum pulse current/kA	200	200	10
Highest pulse frequency/Hz	20	10	1000
Maximum average powe/kW	10	50	30
Minimum pulse duration/μs	30	10	60
Maximum single pulse energy/kJ	36	65	375

Applied various types of electric current pulse, their effects on the solidification process have been conducted on the alloy systems such as pure Al, HT150 cast iron, 1Cr18Ni9Ti stainless steel, Al-Cu and Al-Si etc. Moreover, their effects on different stages of solidification process or pure Al was investigated, and the results showed that the crystal rain caused the structure refinement.

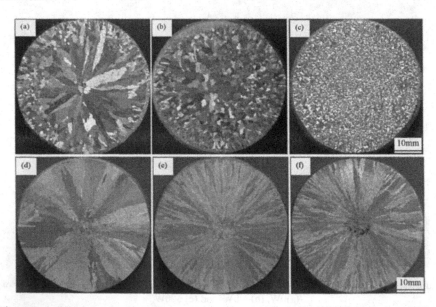

Figure 8. Solidification structure of pure Al (a), (b) and (c) and 1Cr18Ni9Ti stainless steel (d), (e) and (f) under the influence of electric current pulse.

Figure 8 shows that for pure Al, totally refined equi-axed grains could be obtained under the influence of electric current pulse(Figure 8 (b) and (c). However, even strong electric current pulse was employed, yet only thin columnar grains was found, implying that it is more difficult to gain the refined solidification structure in high-melting alloys than that in low-melting alloys.

• Besides, the influence of the electric current pulse on the liquid-solid interface stability was investigated (shown in Figure 9), indicating that the stability was decreased with increasing the peak value of the electric current pulse.

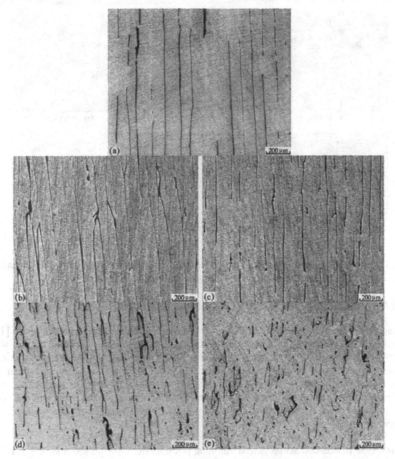

(a) Untreated (b) I=10A (c) I=20A (d) I=30A (e) I=40A
Figure 9. Microstructure of Al-4.5wt%Cu under different peak value of the electric current pulse.

Metals solidification under the influence of pulsed magnetic field [18]

Coupled with electromagnetic coil and the aforementioned pulse power sources, different types of pulsed magnetic fields were available to conduct their influences on solidification process of metallic systems. The magnetic intensity depended on the diameter of the electromagnetic coil, and the highest intensity arrived 10T. Pure Al, HT150 cast iron, 1Cr18Ni9Ti stainless steel were chosen. The results showed that totally equiaxed grains gained in pure Al, however, only refined columnar crystals were gained in 1Cr18Ni9Ti system, yet no equiaxed grains produced.

Figure 10 shows the effect of magnetic intensity on stability of liquid-solid interface of 1Cr18Ni9Ti stainless steel. It was found that the interface stability was largely decreased with increasing the magnetic intensity, owing to the intense perturbation resulting from the pulsed magnetic field.

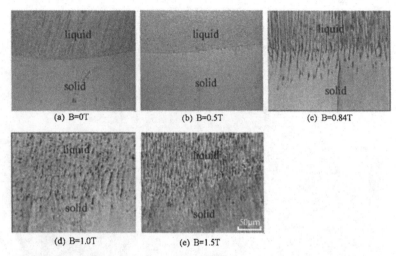

(a) B=0T   (b) B=0.5T   (c) B=0.84T

(d) B=1.0T   (e) B=1.5T

Figure 10. Effect of magnetic intensity on liquid-solid interface stability of 1Cr18Ni9Ti stainless steel at the growth rate 12μm/s (G/V=5.23×10^3Ks/mm^2).

Metals solidification under the influence of magnetic oscillation

The magnetic oscillation was the recently developed external field by our research group [19]. If a pulsed magnetic field is applied along the surface of liquid metals, an induced current will be generated perpendicular to the magnetic field direction. The interaction between magnetic field and current gives rise to the occurrence of internal Lorentz force normal to the surface of liquid metals. Due to the small pulse width of magnetic field the width and depth of induced current are very small. Consequently, no electric or magnetic field occurs inside the melt, and the Lorentz force only acts on the surface of liquid metals yet the magneto oscillation can transport to the interiors.

As a new type of non-contacting pulsed vibrating physical field, the pulsed magnetic oscillating has the features analogous to the ultrasonic with a slight difference of lacking amplitude transformer horn. Meanwhile, the pulsed magnetic oscillating also has the characteristics of electric current pulse without electrodes. Furthermore, the presence of non-contacting advantage of pulsed magnetic field protects the melt from splash and slag entrapment. As a consequence, the pulsed magnetic oscillating provides a novel method not only in the investigations of high temperature metals but also in the study on refinement mechanism with pulsed physical field. Further work will be undertaken to study the effect of such physical field on the solidification process of liquid metals.

Summary and prospect

(1) Extensive investigations have been conducted on the influence of external fields on materials processing in Shanghai University in last decade, and the research fields involve nonferrous and ferrous metals.
(2) External fields can significantly impose on the materials processing, refine the solidification structure of metals, and change their growth orientation.
(3) Attributing to intrinsic characteristics of non-polluting and easy manipulation, the external fields such as electric current pulse and magnetic oscillation are anticipated to be applied in the metallurgy industry in the near future to improve the quality of casting billets.

Acknowledgement

The authors gratefully acknowledge the financial supports from the National Natural Science Foundation of China (Grant No. 50274050, 50374046 and 50574056) and the Pre-research Foundation of the National Basic Research Program (Grant No. 2004CCA07000).

References

1. R.W. Smith, "Electric Field Freezing," *J. Mater. Sci. Lett.*, 6 (1987), 643-644.
2. T.J.Li , and J.Z. Jin, "Current Status and Further Potential for Electromagnetic Processing of Materials," *Mater. Rev.*, 14 (2000), 12-19.
3. O.V. Abramov, "Action of High Intensity Ultrasonic on Solidifying Metals," *Ultrasonics*, 25 (1987), 73-76.
4. X. Jian et al., "Effect of Power Ultrasound on Solidification of Aluminum A356 Alloy," *Mater. Lett.*, 59 (2005), 190-193.
5. Z.S. Lei et al., "Experimental Study on Mould Oscillation-less Continuous Casting Process under High Frequency Amplitude-Modulated Magnetic Field," *ISIJ Int.*, 44 (2004), 1842-1846.
6. H.F. Dong, Z.M. Ren, and K. Dong, "Basic Study of Initial Solidification in Soft Contact Electromagnetic Continuous Casting." *J. Shanghai University (Natural Science)*, 3 (1997), 179-182.
7. H.F. Dong, Z.M. Ren, and Y. B. Zhong, "Homogenization of the Magnetic Field in Soft Contact Electromagnetic Continuous Casting," *J. Iron Steel Res.*, 10 (1998), 5-8.
8. Z.M. Ren et al., "The Segregated Structure of MnBi in Bi-Mn Alloy Solidified under a High Magnetic Field," *Mater. Lett.*, 58 (2004), 3405-3409.
9. Y.S. Liu et al., "Effect of Magnetic Field on the TC and Magnetic Properties for the Aligned MnBi Compound," *Solid State Comm.*, 138 (2006), 104-109.
10. X. Li, Z.M. Ren, and Y. Fautrelle, "Effect of a Vertical Magnetic Field on the Dendrite Morphology during Bridgman Crystal Growth of Al-4.5wt%Cu," *J. Cryst. Growth*, 290 (2006), 571-575.
11. H.B. Zhang, Q.J. Zhai, and F.P. Qi, "Effect of Side Transmission of Power Ultrasonic on Structure of AZ81 Magnesium Alloy," *Trans. Nonferrous Met. Soc. China*, 14 (2004), 302-305.
12. F.P. Qi et al., "Microstructure Refinement of Sn-Sb Peritectic Alloy under High-Intensity Ultrasound Treatment," *Journal of Shanghai University (English Edition)*, 9 (2005), 74-77.

13. Z.Q. Li, F.P. Qi, and Y.Y. Gong, "Effects of Melt Temperature on Solidified Structure of Sn-Sb Peritectic Alloy in Ultrasonic Field," *Special Casting and Nonferrous Alloys*, 3 (2003), 21-22.
14. F.P. Qi et al., "Effect of Ultrasonic Power on Solidification Structure of Sn-Sb Alloy," Journal of Shanghai University (Natural Science), 9 (2003), 43-46.
15. J.H. Fan et al., "Effects of Pulse Electric Current with Various Parameters on Solidification Structure of Austenitic Stainless Steel," *Foundry Technology*, 24 (2003), 534-536.
16. Y. Chen et al., "Effect of High Frequency Pulse Electric Current on Solidification Structure of HT150 Gray Cast Iron," *Foundry*, 53 (2004), 611-613.
17. C.H. Gan, "Effect of Pulse Current on the Solidification Structure of Hypoeutectic Al-Si Alloy," *Foundry Technology*, 27 (2006), 252-254.
18. Q.S. Li, L.Q. Liu, and Q.J. Zhai, "Analysis of Graphite Morphology of Gray Cast Iron in Pulse Magnetic Field," *Journal of Iron and Steel Research*, 12 (2005), 45-48.
19. Q.J. Zhai et al., "Process and Apparatus of Magneto Oscillation for Solidification Structure Refinement of Metals," *Chinese patent*. Application no. 200510030736.4.

EXPERIMENTAL STUDY ON SOLIDIFICATION STRUCTURE OF AL

INGOT BY PULSE MAGNETO OSCILLATION

Yong-Yong GONG, Jin-Xian JING, Chang-Jiang SONG, Yu-Lai GAO, Qi-Jie ZHAI
Center for Advanced Solidification Technology (CAST), Shanghai University;
Shanghai 200072, P.R. China

Keywords: Pulse magneto oscillation, Grain refinement, Solidification structure.

Abstract

In this paper, a novel technique for grain refinement was proposed by using pulse magnetic field which was named as Pulse Magneto Oscillation (PMO). The PMO method possesses the advantages, such as non-contacting, no melt splashing, and convenient operation. Investigation results show that PMO method can significantly refine the solidification structure of Al ingot. And there is an optimal current intensity for structure refinement. Below it, increasing the current intensity can make the structure become finer.

Introduction

During the melt solidification, the application of physical field is an important microstructure control technology [1]. The physical fields generally include electric current[2~5], magnetic field[6~9], ultrasonic waves, and slight gravitational field etc. The actions of electromagnetic field on the liquid metal include: Electrics transmission, Joule effect, Peltier effect, Lorentz force, and inoculation etc. Since 1960, applications of electromagnetic field to melts solidification have been extensively studied, and great achievement was acquired in refining of the metals solidification structure.

In this paper, a novel grain refinement method by a novel pulse magnetic field was proposed, which was named as Pulse Magneto Oscillation (PMO) [10]. Its principle is that when a narrow pulse magnetic field parallelly acts on the surface of melt, it will bring induction current in the melt, the induction current and the pulse magnetic field will produce Lorentz force and the vibration will be induced in the melt. This technique possesses the advantages, such as no melt splashing like ultrasonic and convenient operation of magnetic field which provides a good method to grain refinement.

Experiments

Figure1 and Figure 2 schematically show the experimental setup and the power of PMO. The industrial pure aluminum(≥99.7wts.%) was used as the studied metal. It was melt and heated to 750℃ by an electric resistance furnace and was poured to a ceramics mould in size of Φ40mm×150mm at 700℃. During the solidification of liquid Al, it was treated by the PMO. The parameter of PMO includes the magnetic-field strength B or induced current strength J, pulse width τ, and pulse frequency f.

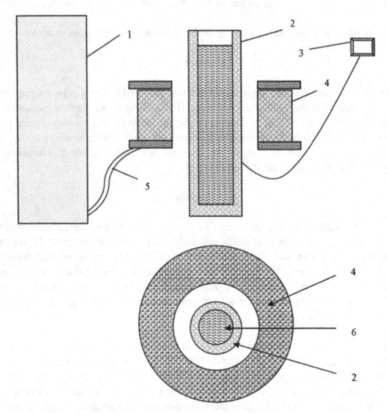

Figure 1. The illustration of the experiment set-up: 1. Pulse current power, 2. Ceramic mould and melt, 3. Temperature measurement system, 4. PMO generator, 5. Wire, 6. Melt.

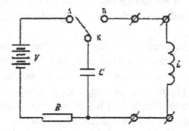

Figure 2. The schematic of the circuit of PMO.

Experimental results

The solidification of liquid Al was treated by PMO with higher charged voltage and lower frequency (1.2Hz). The longitudinal structure of the sample are shown in Figures 3 (b)-(e). The solidification structure without PMO is large columnar grain in the most part of the sample and a few equiaxed grains at the bottom in addition a deep shrinking hole appears in the sample.

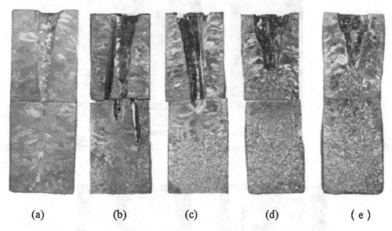

     (a)   (b)   (c)   (d)   (e)

Figure 3. Longitudinal macrostructure of the samples treated under different discharge voltages with a frequency of 1.2Hz: (a) untreated, (b) 800V, (c) 1600V, (d) 2400V, (e) 3200V.

Figure3 (b) is the longitudinal structure of the sample treated by PMO at 800V. In it the number of equiaxed grains increases, and the grains are finer than that untreated sample at the bottom of the sample. The longitudinal structure of sample treated by PMO at 1600V is shown in Figure 3(c). Its structure is clearly different from untreated sample, the number of equiaxed grains greatly increases, and the grains are much finer than that of untreated sample. The longitudinal structure of sample treated by PMO at 2400V are almost uniform fine equiaxed grains except some larger grains near the shrinking hole at the top of the sample and the depth of the shrinking hole decreases. The structure of the sample treated by PMO at 3200V is nearly the same with that of the sample (d). But its refinement effect is not as good as sample (d).

Figure 4 is the longitudinal structure of the sample with the transverse stainless steel mesh: Figure 4 (a) is without PMO treatment, and Figure 4 (b) is treated by PMO whose coil located above the stainless steel mesh. Results show that there are fine grains only above the stainless steel mesh in Figure 4(b). Figure 5 is the longitudinal structure of the samples with the longitudinal stainless steel mesh: Figure 5(a) is without PMO treatment, and Figure 5 (b) is treated by PMO. In Figure 5 (b), the sample was refined only out the stainless steel mesh.

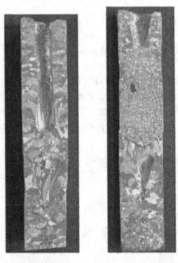

(a) untreated   (b0 treated

Figure 4. Effect of the transverse stainless steel mesh on structure refinement by PMO.

(a) untreated   (b) treated

Figure 5. Effect of the longitudinal stainless steel mesh on structure refinement by PMO.

Discussion

From the experimental results, it can be seen that with increasing the current intensity, the structure is markedly improved, equiaxed grains area gradually increases, columnar crystal area gradually decreases, the size of equiaxed and columnar crystal decreases gradually. When the current intensity of PMO is very small, the structure refinement is not obvious as shown in Figure 3(b). With the increase of the current intensity, the structure gradually become more and more finer as shown in Figure 3(c)~(d). When the current density exceeds a critical value, refinement effect begin to deteriorate as shown in Figure 3 (e). This suggests that there is an optimal current intensity for structure refinement.

PMO microstructural refinement mechanism may be that during the solidification process of the liquid metal, the solid phase particles firstly solidified close to the mould wall migrate central-ward by the electromagnetic force and/or shockwave induced by the pulsed magnetic field and fall down to the bottom due to their gravity, acting as the crystal nucleus and making the structure refine.

Conclusions

In aluminum ingot, PMO can make area of the equiaxed grains increase, area of the columnar crystals decrease and the solidification structure significantly refine. There is an optimal current intensity for structure refinement. Below it, increasing the current intensity makes the structure finer. And its refinement mechanism may be that the solid phase particles firstly solidified close to the mould wall migrate central-ward and downward by the coupling action of the pulsed magnetic field and gravity.

Acknowledgement

The authors gratefully acknowledge the financial supports from the National Natural Science Foundation of China (Grant No.50574056)

References

[1] FAN Jin-hui, and ZHAI Qi-jie, "Effects of physical fields on solidification structure of metals," The Chinese Journal of Nonferrous Metals, 12(2002), 11-17.
[2] LI Hui, BIAN Xiu-fang, and LIU Xiang-fa, et al.,"Influence of current density on structure and properties of Al alloy," Special Casting and Nonferrous Alloys, 3(1996), 8-10.
[3] Nakada M, and Flemings M G, "Modification of solidification structures by pulse electric discharging," ISIJ International, 30(1990), 27-33.
[4] Barnak J P, Sprecher A F, and Conrad H, "Colony (grain) size reduction in eutectic Pb-Sn castings by electroplusing," Scripta Metallurgica et Materialia, 32(1995) , 879-884.

[5] LI Jian-ming, and LIN Han-tong, "Modification of solidification structure by pulse electric discharging," Scripta Metallurgica et Materialia, 31(1994), 1691-1694.

[6] NING Zhi-liang, XU Zhi-hui, and LIANG Wei-zhong et al., "Effect of pulsating electromagnetic field on hot cracking of Al-Cu alloys," The Chinese Journal of Nonferrous Metals, 7(1997), 129-133.

[7] ZHANG Qin, and CUI Jian-zhong, "Microstructural refinement mechanism on aluminum alloy produced by semi-continuous casting in electromagnetic vibration," The Chinese Journal of Nonferrous Metals, 13(2003), 1500-1504.

[8] BAN Chun-yan, CUI Jian-zhong, and BA Qi-xian et al., "Influence of pulsed current and pulsed magnetic field on distribution of alloying elements in Al alloy," The Chinese Journal of Nonferrous Metals, Al Special(2002), 121-123.

[9] Hans C, "Influence of an electric or magnetic field on the liquid-solid transformation in materials and on the microstructure of the solid," Materials Science and Engineering A, 287(2000), 205-212.

[10] Q. J. Zhai, Y. Y Gong, and Y. L. Gao et al., Process and apparatus of magneto oscillation for solidification structure refinement of metals, Chinese patent, No: 200510030736.4.

Materials Processing under the Influence of External Fields
Edited by Qingyou Han, Gerard Ludtka, Qijie Zhai
TMS (The Minerals, Metals & Materials Society), 2007

EFFECT OF PULSED MAGNETIC FIELD ON THE SOLIDIFIED STRUCTURE OF PURE AL MELT

Yu-Lai Gao[1], Qiu-Shu Li [1, 2], Hai-Bin Li[1,2], Qi-Jie Zhai[1]

[1]Center for Advanced Solidification Technology (CAST), Shanghai University; Yanchang Road
149, ZhaBei District; Shanghai, 200072, P.R. China
[2]School of Materials Science and Engineering, Taiyuan University of Science & Technology,
Taiyuan 030024, P.R. China

Keywords: Pulsed magnetic field, Structure refinement, Inoculation, Nucleation and growth.

Abstract

A comparative study was made of the effect of pulsed magnetic field on the solidification structure of pure aluminum by employing the self-developed high voltage pulsed power source at several solidification stages. It has been shown that the solidified structure was remarkably refined with applied pulsed magnetic field during or in the overall solidification process. In contrast, no significant difference in the solidification structure of aluminum was observed also with pulsed magnetic field prior to the stage of nucleation. As a result, it was convincing that the structure refinement arises from the pulsed magnetic field at the stage of nucleation and growth, yet the inoculation effect was negligible.

Introduction

Magnetic field plays a particular role in materials processing, especially in the field of metal solidification. At present, several types of magnetic fields including direct current, alternating current, travelling wave, rotating and pulsed magnetic fields were employed to refine the solidification microstructure, which, in essence, differs from the traditional inoculation treatment due to their protecting the liquid metal from being polluted. It is, therefore, a new type of solidification structure refinement technique.

Direct current magnetic field can efficiently impair the convection during solidification and improve the solidification microstructure[1-3], whereas the aim of applying alternating and rotating magnetic fields is to generate electromagnetic stirring [4, 5] or levitation[6, 7], altering the solidification process and controlling the solidification microstructure.

Different from the said types of magnetic fields, pulsed magnetic field processing has become one of the most promising new techniques to refine the solidified structure attributing to its large energy capacity and easy manipulation. Vives[8, 9], Zi [10] and Asai[11] et al. investigated the effect of different magnetic fields on the solidified structure of metal materials, indicating that the solidified structure can be effectively refined by introducing magnetic field. Bassyouni[12] pointed out that optimal refinement can be obtained only the intensity of the magnetic induction intensity is in the range of 0.027~0.033 T. However, Ge [13] thought that the refinement effect not only depended on the magnetic intensity, but also determined by the intensity and waveshape of the primary current together with the frequency of the pulse, and an evident refinement was achieved when the magnetic induction intensity is 0.2T. As a result, no consensus is acquired about the effect of magnetic field on the solidified structure of metals. Also, little information is available about the refinement mechanism. Many researchers realized that the solidified structure was controlled by several factors, including the properties and parameters of the magnetic field, as well as the properties and solidification conditions of materials. Consequently, it remains an urgent problem to conduct an in-depth study on the basic research in order to explain the effect of magnetic field on the solidified structure.

Experimental procedure

The experiment was carried out by using the self-designed high-voltage pulsed power source and the solidification tester. Figure1 shows the schematic diagram of the solidification tester consisting of a high-voltage pulsed power source over 10^4V, a generator to produce intense electromagnetic field, a computer-controlled temperature acquisition system, a cooling system and a mobile platform etc.

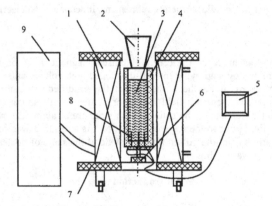

Figure 1. Schematic diagram of the experimental tester for solidification under high-voltage pulsed magnetic field.
1-Magnetic field generator, 2-Pouring cup, 3-Liquid metal, 4-Mould, 5-Computer-controlled thermometric system, 6-Pad, 7-Mobile platform, 8-Thermal couples, 9-High-voltage power source

The size of the generator chamber to produce intense magnetic field is Φ75mm. The highest value of the intensity of pulsed magnetic field can reach 6T in the central homogeneous zone. The real-time temperature measurement was carried out by the computer-controlled temperature collection system. The thermal couple was inserted from the bottom of the foundry mould, and the depth of the penetration was 30mm.

The distribution of the magnetic field inside the generator in the axial and radial directions was measured to ensure that the similar magnetic intensity remains during the solidification process (shown in Figure 2 and Figure 3). It is clearly that a homogeneously distributed magnetic zone is available at the center of the generator.

To shed more light on the effect of pulsed magnetic field on the metal materials, pure aluminum was chosen as an example because of its single component and the solute segregation in equilibrium and non-equilibrium conditions are not expected, which is favorable to avoid the influence of solute on the mechanism of magnetic field.

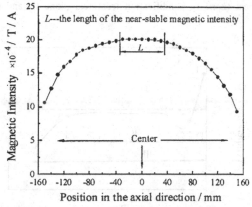

Figure 2. Distribution of the magnetic intensity in the axial direction.

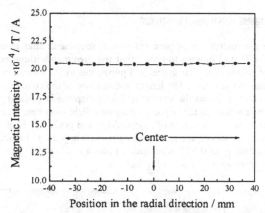

Figure 3. Distribution of the magnetic intensity in the radial direction.

The aluminum ingots were placed in a graphite crucible, and melted in an electrical resistance furnace. In view of the temperature decrease, the liquid metal was superheated to 850°C and subsequently poured into the mould. The foundry mould was presented in the chamber of the generator (Figure 1). The size of the samples is Φ50×130mm. The cross-section of the treated specimens was gained by being cut from the lower part of the sample, which is 10mm away from the top of the thermal couple.

Results and discussion

Figure 4 shows the typical cooling curve of pure Al melt. It is known that it undergoes three stages during the solidification process of pure Al melt, namely liquid cooling stage (point 1~2), solidification stage (points 2~3) and solid cooling stage (after point 3). To study the influence of the pulsed magnetic field on the solidification of pure Al, the external field was employed at various solidification stages, indicated as zone I, II, and III in Figure 4. These zones were defined as inoculation treatment (zone I), solidification treatment (zone II) and overall solidification treatment (zone III).

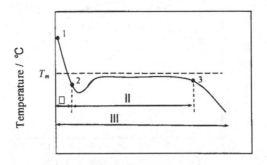

Time / s

Figure 4. Typical cooling curve of pure Al melt.

Solidified structure under inoculation treatment

Figure 5 shows the macrostructure of pure aluminum specimens after incubation treatment of pulsed magnetic field. It is of interest to notice that the specimen without pulsed magnetic field treatment consists of coarse columnar grains and penetrates to the interior with small amount of coarse equiaxed grains in the center. The length and amount of the columnar grains was reduced and distributed at random, whereas the amount of finely dispersed equiaxed grains was increased when the specimen was subjected to the pulsed magnetic field treatment with varying intensities. Irrespective of the fact that the variation of morphology and grain size is obvious with increased magnetic field intensity, however, the final macrostructure appears to be undesirable even if the high magnetic field intensity of 0.51T is reached in that a small quantity of columnar grains are present in the macrostructure, as shown in Figure 5(f).

Figure 5. Macrostructure of pure Al under inoculation treatment
(a) 0T (b) 0.12T (c) 0.22T (d) 0.31T (e) 0.37T (f) 0.51T

64

Further increase in magnetic field intensity may lead to the much refiner macrostructure. At this moment, however, the metal melt vibrates intensively and splashes out of the mold, making it difficult to perform the experiment. This indicates that a limited but not optimal incubation effect can be obtained as the pulsed magnetic field intensity exceeds 0.37T.

Solidified structure under solidification treatment

The temperature-time curve was monitored through a computer data-collecting system. And the pulsed magnetic field was applied to treat the Al melt immediately after the cooling process was finished and the inflexion point occurred. The experimental parameters are the same as those of inoculation treatment by pulsed magnetic field.

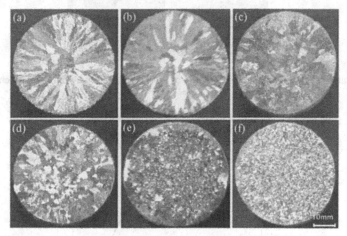

Figure 6. Macrostructure of pure Al under solidification treatment
(a) 0T (b) 0.12T (c) 0.22T (d) 0.31T (e) 0.37T (f) 0.51T.

Figure 6 shows the morphologies of solidified pure aluminum specimens treated by the pulsed magnetic field. It is seen that the macrostructure evolution after pulsed magnetic field treatment is consistent with the case of inoculation treatment. As the pulsed magnetic intensity is increased to 0.31T, the columnar grains are substituted by the equiaxed grains, as shown in Figure 6 (d). Further increase in magnetic field intensity induces a finer and uniformly distributed equiaxed-grained macrostructure throughout the cross section, as depicted in Figure 6(e) and (f).

In order to understand the influence range of pulsed magnetic field in the height direction of specimens, the cross-sectional macrostructures of specimens shown in Figure 7 were analyzed. It is observed that the overall morphology in Figure 7 is in accordance with that in Figure 6. The equiaxed grains are uniformly distributed and macrostructure unioformity is increased with increasing the pulsed magentic field intenstiy. This indicates that range of magnetic field of the self-developed pulsed magnetic field soldification apparatus is wide. The solidified structure of specimen can be refined along the full height of specimen.

The effect of magnetic field intensity on the average grain size during pulsed magnetic field solidification treatment was analyzed by use of metallographic technique, as shown in Figure 8. In spite of the fact that the average grain size decreases with increasing the magnetic field intensity, however, it seems that there is an optimal value of magnetic field intensity to refine the solidified macrostructure from the analysis of the curve. By fitting the curve we obtain the equation, which describes the dependence of mean grain size with pulsed magnetic field.

$$\Psi = 6.59 - 21.71B + 21.55B^2 \tag{1}$$

where Ψ and B are the mean grain size and pulsed magnetic field intensity, respectively.

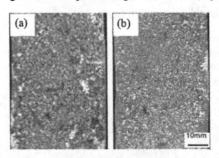

Figure 7. Cross section macrostructure of pure Al under solidification treatment
(a) 0.37T   (b) 0.51T.

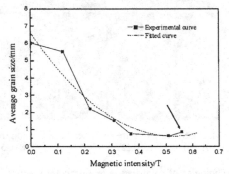

Figure 8. Relation between the magnetic intensity and the average grain size under the solidification treatment.

Differentiating Eq. (1) and let $\dfrac{d\Psi}{dB} = 0$, we arrived

$$B = 0.53T \tag{2}$$

This implies that there exists an optimal pulsed magnetic field to refine the solidification macrostructure. In other words, an optimal refined macrostructure would be anticipated when the magnetic field intensity reaches an optimal value, over which the influence of refinement would probably be tampered. Under the present experimental conditions, the smallest mean grain size was obtained when the magnetic field intensity reaches 0.53T. To validate the aforementioned result, an additional experiment was carried out. Firstly, a magnetic field of 0.56T was selected to treat the liquid metal and the solidified macrostructure with an average grain size of 0.84mm in Figure 9 was obtained. Compared with the solidified macrostructure in Figure 6(f), the structure in Figure 9 is coarser and contains more defects, indicating that high magnetic field intensity is not always beneficial to grain refinement.

Figure 9. Macrostructure of pure Al under the solidification treatment (0.56T).

Solidified structure under overall solidification treatment

Figure 10 shows the solidification macrostructures of samples treated by pulsed magnetic field. It is clearly seen that the variation of morphologies with magnetic field intensity is in accord with the pulsed magnetic field solidification treatment process. The morphology of samples develops from columnar-grained to equiaxed-grained. However, the grain under the same magnetic field intensity is more refiner.

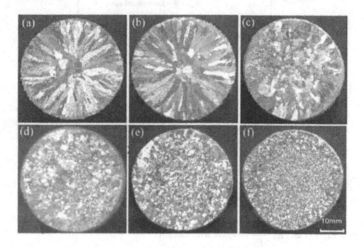

Figure 10. Macrostructure of pure Al under overall solidification treatment
(a) 0T (b) 0.12T (c) 0.22T (d) 0.31T (e) 0.37T (f) 0.51T.

The cross-sectional macrostructure of specimens through magnetic field treatment is shown in Figure 11. In addition to very small amount of coarser equiaxed grains at the edges, the overall section exhibits finely distributed equiaxed grains. Higher magnetic field intensity goes with the reduced grain size.

An analysis was made of the influence of magentic field intensity on the average grain size during the pulsed magnetic field treatment, as depicted in Figure 12. Fitting the original curve leads to the magnetic intenstity dependence of mean grain size

$$\Psi = 6.53 - 23.22B + 22.13B^2 \tag{3}$$

where Ψ is the mean grain size and B is the pulsed magnetic field intensity.

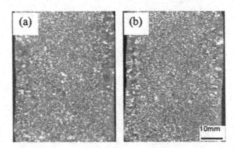

Figure 11. Cross section macrostructure of pure Al under overall solidification treatment
(a) 0.37T (b) 0.51T.

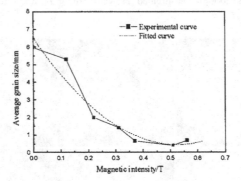

Figure 12. Relation between the magnetic intensity and the average grain size under the overall
solidification treatment.

Differentiating Eq. (3) and let $\dfrac{d\Psi}{dB} = 0$, we also obtain

$$B = 0.52T \tag{4}$$

Again, an additional experiment was performed and the macrostructure was shown Figure 13 with an average grain size of 0.68 mm. In comparison with Figure 10(f) the macrostructure is more coarser, similar to the case of pulsed magnetic field treatment.

The above-mentioned experimental results for pure aluminum melt suggest that the effect of pulsed magnetic field on grain refinement is not obvious prior to the solidification. The solidified macrostructure is markedly refined as the solidification occurs. Due to the characteristic of pure aluminum only equiaxed or columnar grains without dendritics will be formed during growth. It is, therefore, not possible to achieve grain multiplication via grain fragmentation. Thus, it is speculated that the promotion of nucleation during solidification is responsible for the grain refinement.

Figure 13. Macrostructure of pure Al under the overall solidification treatment (0.56T).

Conclusions

(1) The solidification structure of pure Al can be obviously refined under the pulsed magnetic field, and there exists an optimal magnetic intensity to obtain the finest structure.
(2) Various refined structures can be gained when pure Al melt was treated at different solidification stages. Moreover, it was deemed that the structure refinement arises from the pulsed magnetic field at the stage of nucleation and growth, yet the inoculation effect was negligible.
(3) It was thought that the structure refinement was dominated by the promotion of nucleation during the solidification process of pure Al.

Acknowledgement

The authors gratefully acknowledge the financial support from the National Natural Science Foundation of China (Grant No. 50274050).

References

1. C. Hans, "Influence of an Electric or Magnetic Field on the Liquid Solid Transformation in Materials and the Microstructure of the Solid," *Mater. Sci. Eng. A*, 287 (2000), 205-212.
2. D.R. Uhlmann, T.P. Seward, and B. Chalmers, "The Effect of Magnetic Fields on the Structure of Metal Alloy Castings," *Trans. Metall. Soc. AIME*, 236 (1966), 527-531.
3. S. Asai, "Effects of Electromagnetic Forces on Solidified Structure of Metal," *Trans. ISIJ*, 18 (1978), 754-760.
4. A. Ohno, T. Moteg, and T. Shimizu, "Formation of Equiaxed Crystals in Ammonium Chloride-Water Model and Al Alloy Ingots by Electromagnetic Stirring," *J. Japan Inst. Metals*, 46 (1982), 554-563.
5. K. Murakami, T. Fujiyama, and A. Koike, "Influence of Melt Flow on the Growth Directions of Columnar Grains and Columnar Dendrites," *Acta Metall.*, 31 (1983), 1425-1432.
6. R.S. Lin, and M.G. Frohberg, "Determination of Mixing Enthalpies of the Liquid Vanadium-Silicon System by Levitation-Alloying Calorimetry," *High Temp.-High Press*, 24 (1992), 543-550.
7. M. Barth, B.B. Wei, and D.M. Herlach, "Dendritic Growth Velocities of the Intermetallic Compounds Ni2TiAl, NiTi, Ni3Sn, Ni3Sn2 and FeAl," *Mater. Sci. Eng. A*, 226 (1997), 770-773.

8. C. Vives, "Effects of Forced Electromagnetic Vibrations during the Solidification of Aluminum Alloys.1. Solidification in the Presence of Crossed Alternating Electric Fields and Stationary Magnetic Fields," *Metall. Mater. Trans. B*, 27 (1996), 445-455.

9. C. Vives, "Crystallization of Aluminum Alloys in the Presence of Vertical Electromagnetic Force Fields," *J. Cryst. Growth*, 173 (1997), 541-549.

10. B.T. Zi, Q.X. Ba, and J.Z. Cui, "Effect of Strong Pulsed Electromagnetic Field on Metal's Solidified Structure," *Acta Physica Sinica*, 49 (2000), 1010-1014.

11. S. Asai, and Birthand, "Recent Activities of Electromagnetic Processing of Materials," *ISIJ Int.*, 29 (1989), 981-992.

12. T.A. Bassyouni, "Effect of Electromagnetic Forces on Aluminum Cast Structure," *Light Metal*, 33 (1983), 733-742.

13. F.D. Ge, H.L. He, and S.X. Huo, "Casting with Pulsating Electromagnetic Field," *Chin. J. Mech. Eng.*, 25 (1989), 1-6.

Materials Processing under the Influence of External Fields
Edited by Qingyou Han, Gerard Ludtka, Qijie Zhai
TMS (The Minerals, Metals & Materials Society), 2007

HIGH MAGNETIC FIELD INFLUENCES ON FABRICATION OF MATEIRALS WITH MAGNETIC PHASES

K. Han and B. Z. Cui

National High Magnetic Field Laboratory, Florida State University, Tallahassee, FL 32310

Key words: Magnetic annealing, Magnetic properties, Microstructure Refinement

Abstract

This paper discusses the effects and relevant mechanism of high magnetic fields on fabrication of materials. External magnetic fields influence either the texture or microstructure of the materials. The external fields change the microstructure by adding extra terms to the Gibbs free energy so that the phase transformation temperatures or the driving force for the phase transformation is altered. By changing the phase transformation temperatures, the magnetic field can accelerate or delay the phase transformation kinetics and increase or decrease the nucleation site numbers for the formation of magnetic phases. At the same time, the external magnetic field can change the growth rate. Increase of the nucleation rate results in a refinement of the microstructure and improves the properties that require fine structured materials. On the other hand, magnetic annealing in a high field is an emerging method for manufacturing textured materials. Consequently, the material properties can be improved by magnetic-field-induced crystallographic textures and phase transformations.

Introduction

The National High Magnetic Field Laboratory (NHMFL) in USA builds and operates various high field magnets for research. One of the important research areas is to use high magnetic fields to engineer the microstructure and texture of materials so that the properties of the materials can be improved.

Magnetic annealing has been given special attention in the microstructural refinement and materials texturing to enhance the material properties. For example, magnetic properties for the magnetic materials can be enhanced by a high magnetic field annealing (HMFA). The HMFA is more powerful if it is combined with other techniques. Severe plastic deformation and subsequent HMFA was found to be able to offer a promising novel way to fabricate bulk, fully dense, anisotropic magnetic nanocomposite with a record-high energy product $(BH)_{max}$. Majority of the previous magnetic field annealing experiments have been done on materials with relatively low fields due to the limitation of the field strength. With the increase of the magnetic field strength and magnet bore sizes, the HMFA on large size of samples is possible and materials with better properties can be fabricated by the HMFA. Therefore, it is useful to study the high magnetic field impacts on fabrication of materials with optimized properties. In addition, the mechanisms through which the HMFA plays an important role on textures, microstructures and magnetic properties require further explorations [1]. This paper outlines some of the researches undertaken in these areas by different groups and addresses some issues required to be further studied. Particularly, the paper focuses on the effects and relevant mechanism of HMFA on nanostructures, crystallographic textures, exchange coupling, magnetic anisotropy, and hard magnetic properties for nanostructured permanent magnets.

Magnetic Field Effects on Phase Transformations

It has been found that the external high magnetic field promotes first order phase transformation. Earlier work concentrated on the systems where phase transformation is related to form ferromagnetic phases. Many researchers reported that the external magnetic fields introduce diffusionless phase transformation in various steels and alloys [2, 3, 4, 5, 6, 7, 8, 9, 10, 11, 12, 13, 14, 15,]. Typical phase transformation is to form martensite. It was argued that martensite starting transformation temperature could be raised up to 3 decrees per Tesla (T) [2,3,4] in steels. On the other hands, the high magnetic field has also effects on diffusion-controlled phase transformation, such as bainite phase transformation and pearlite phase transformation. In high carbon steels, a magnetic field of 1.2 T increases the amount of hypoeutectic ferrite during the transformation of autenite into ferrite and carbides [16]. In pearlite, ferrite shows supersaturated carbon content when the phase transformation occurs at 560°C with field gradient of 50 T/m [17]. However, it appears that magnetic fields have no influence on steels without an addition of Cr and Mn. Therefore, the existence of the transition metals is important in steel fabrications in high magnetic fields. Although it is expected that high magnetic fields have more influence on ferromagnetic phases, it is realized that the magnetic fields can accelerate the phase transformation from austenite to ferrite both below and above the Curie temperatures in Fe-C system [18]. Therefore, the high magnetic fields appear to have impact on phase transformation in various of phases.

Other cases were reviewed recently [19]. The study of the magnetic field induced phase transformation is not only useful in identify the critical issues in basic science study on magnetic field influence on phase transformations, it is also useful in application of the materials, such as the development of structural materials with high mechanical strength, functional materials with high energy products for magnet applications, superconducting materials with high critical current.

The magnetic field influence on phase transformation can be calculated by classic thermodynamics. Many groups have used such approach and most recent one examined the effects of applied magnetic fields on the α-bcc/γ-fcc phase boundary in the Fe-Si system in the paramagnetic state. The contributions to the total Gibbs energy of the α-bcc and γ-fcc from the external fields are calculated based on the Curie-Weiss law and susceptibility data. The Fe-Si phase diagram on the Fe-rich side as a function of applied field is studied using the Thermo-Calc (TC) package. With increase of the magnetic field strength, the α-bcc and γ-fcc transition temperature increases while that for γ-fcc /δ-bcc transition decreases. This indicates that the bcc phase becomes stable with the external magnetic field. However, the field strength has less influence on the phase boundary, γ-fcc /δ-bcc, at higher transition temperature boundary than on the lower one [20].

The phase boundary changes by high magnetic fields may affect both the nucleation rate and growth rate of new phase formations because of the change of the driving force for nucleation and activation energy for diffusion. Phase transformation is generally considered as a nucleation and growth process. In the nucleation theory, the steady state homogeneous nucleation rate I_{st} is given by [21]

$$I_{st} = I_0 * exp(-Q/RT) * exp(-\Delta G_c/RT) \quad (1),$$

where I_0 is a pre-exponential factor, Q is the activation energy for the transfer of atoms across the surface of the nucleus, which is approximately equal to the diffusion activation energy, and ΔG_c is the free energy required to form a nucleus of the critical size. ΔG_c can be written as

$$\Delta G_c = A\gamma^3/(\Delta G_v^2) \quad (2),$$

where ΔG_v is the Gibbs free energy difference between the new formed crystals and the matrix phase, ie. $\Delta G_v = G_{crystal} - G_{matrix} < 0$. A is a coefficient and γ is the interfacial energy of the crystal/matrix interface. An interesting issue is that the energy difference can also be the results of magnetic field only. In ferromagnetic systems, the potential energy can be lowered by $\mu_0\Delta M(T)_s \bullet H$, if the difference between the saturation magnetization of two phases is $\Delta M(T)_s$ [22]. One can use crystallization of amorphous phase of Nd-Fe-B into $Nd_2Fe_{14}B/\alpha$-Fe as an example. The matrix of amorphous can be fabricated by melt-spinning and the final products are $Nd_2Fe_{14}B$ (2:14:1)/α-Fe-type nanocomposite. $G_{crystal}$ can be the Gibbs free energy $G_{\alpha\text{-Fe}}$ or $G_{2:14:1}$ for the α-Fe and $Nd_2Fe_{14}B$ phases, respectively, or can be the combination of two phases. The amorphous matrix has a Curie temperature T_c of $310^\circ C$. Therefore, the matrix is paramagnetic during the crystallization process from the amorphous matrix. The amorphous material is annealed at $690^\circ C$ and the crystallization occurs. The newly formed nanocrystallines are a ferromagnetic α-Fe phase ($T_c = 771^\circ C$) and a paramagnetic $Nd_2Fe_{14}B$ phase ($T_c = 325^\circ C$). When a magnetic field is applied to the system, the Gibbs free energy $G_{\alpha\text{-Fe}}$ of the α-Fe nanocrystallines decreases due to the addition of the magnetostatic energy. As the new phase α-Fe is still ferromagnetic and the matrix is paramagnetic, the extra Gibbs free energy difference ΔG_H introduced by a magnetic field H is

$$\Delta G_H = -\mu_0\Delta M(T)\bullet H - 1/2*\chi*H^2 \qquad (3)$$

The first term, $-\mu_0\Delta M(T)\bullet H$, represents the magnetostatic energy, where $\Delta M(T)$ is the difference in magnetization between the α-Fe and amorphous matrix at transformation temperature. The second term, $-1/2*\chi*H^2$, represents the energy due to the high field susceptibility, where χ is the high field susceptibility in the matrix. In this case, there is little volume change associated with the phase transformation during the crystallization process. Therefore, the energy term due to forced volume magnetostriction effect described in reference [23] is neglected here.

It follows that the presence of H increases the absolute value of ΔG_v, resulting in an increase of the driving force for crystallization and further, the nucleation rate I_{st} of α-Fe. The above analysis also indicates that the higher the magnetic field the greater the I_{st} for α-Fe is. Therefore, the HMFA promotes crystallization of the amorphous matrix and increases the nucleation site numbers for the formation of α-Fe phase during the crystallization process. It was reported that the nucleation rate of ferromagnetic ferrite was remarkably accelerated by three times under the influence of a 10 T field in the Fe-C alloy. However, the growth rate is nearly the same with and without a 10 T field, so the magnetic field has little effect on the activation energy for the crystal growth of ferrite [24]. In $Nd_2Fe_{14}B/\alpha$-Fe system, the HMFA leads to reduced α-Fe grain sizes and more uniform distribution of the grains for both the hard and soft phases, if one compares the HMFA without the magnetic field. XRD investigations demonstrate that the average grain sizes of α-Fe are 17 nm and 20 nm, respectively, for the melt-spun $Nd_{2.4}Pr_{5.6}Dy_1Fe_{84}Mo_1B_6$ samples annealed at $690^\circ C$ for 20 min with and without an in-plane field of 19 T. The TEM results show that, HMFA introduces finer and more homogeneously distributed soft and hard nanograins. Similar results were also observed in the melt-spun $Nd_2Fe_{14}B/Fe_3B$-type nanocomposites [25].

On the other hands, the Gibbs free energy of the paramagnetic phases, such as the $Nd_2Fe_{14}B$ phase in $Nd_2Fe_{14}B/\alpha$-Fe -type nanocomposites, should also decrease slightly, so the nucleation rate I_{st} of the paramagnetic phases should increase marginally in the presence of a high magnetic field. Thus, the HMFA also promotes crystallization of the amorphous matrix to form the $Nd_2Fe_{14}B$ phase in the paramagnetic state when the applied field reaches as high as 19 T. Additionally, HMFA improves the crystalline quality of the hard $Nd_2Fe_{14}B$ phase. But these changes are believed to be much smaller than those of ferromagnetic phase, such as α-Fe, as the $Nd_2Fe_{14}B$ phase is in a paramagnetic state during the HMFA process.

The refinement of the precipitation due to the phase transformation affected by high magnetic field was also found in other systems. In Mg alloys, the density of the $Mg_{17}Al_{12}$ continuous precipitates inside the grains was increased and the precipitation plates became thinner in high magnetic field pre-aged specimens. The authors attributed the refinement of the precipitates to the retarded volume diffusion by the high magnetic fields [26].

73

In polycrystalline materials, the nucleation of the new phases occurs at the grain boundaries. Therefore, the driving force in equation (1)-(3) should take the grain boundary influences into account. The acceleration of the formation of the grain boundary precipitation was found in the discontinuous precipitation of $Mg_{17}Al_{12}$ at grain boundaries in a 10-T field. This was related to the magnetically induced grain boundary migration [26].

High magnetic field also influences the second order phase transformations. It is found that the ordering rate in magnetron sputtered FePt films is accelarated by applying a magnetic field during the post-deposition annealing [27]. It is known that FePt shows the fcc-fct disorder-order phase transformation at about 450°C when the film was annealed in a magnetic field. At this temperature, the ordered fct FePt phase is ferromagnetic with a high magneocrystalline anisotropy, while the disordered fcc phase becomes paramagnetic, because the Curie temperature T_c for the disordered and ordered FePt are 330 and 480°C, respectively [28]. When a magnetic field is applied at 450°C. The total free energy of the ferromagnetic phase is reduced by ΔG_H, which is thought to be the additional energy of the phase transformation in FePt films. It is observed that there are more nuclei of the ordered FePt in the magnetically annealed samples than that annealed without a field. This means that the HMFA will effectively increase the nucleation rate of the ordered FePt phase and the areal density of the ordered nuclei, and further increase the fcc-fct transformation rate [29].

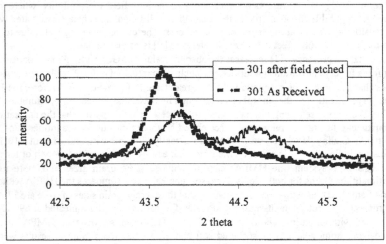

Figure 1. X-Ray data of 111γ (right peak) and 110α (left peak) of an SS 301 before and after the magnetic field treatments.

Although high magnetic field may improve the properties of the materials by influence the phase transformation of the materials, the materials are expected to be stable during the service in high magnetic field. The NHMFL has been using various high strength and high modulus reinforcement materials to achieve high field because the magnetic stresses exceed the strength of most of the conductors. Cold-rolled SS301 stainless steels are one of the candidates as reinforcement materials. The as-rolled SS301 stainless steels are usually composed of martensite and metastable austenite. When the materials are used as reinforcement for magnets, the metastable austenite in the stainless steels may transform into martensite under certain conditions such as (1) in a high magnetic field (2) by further deformation under cyclic loading (3) at cryogenic temperatures. The combination of the two or three factors can results in unexpected changes of the martensite transformation behavior further and it is unclear if those factors have any combination effects on the mechanical properties. Although it was reported that the magnetic field and cryogenic environments promote the phase transformation in SS304, no research has been undertaken on the influence of these conditions on martensite phase transformation in SS301, and the phase transformation will alter the properties of the materials. Therefore, we have undertaken the research on the instability of the SS301 at 77K in magnetic fields. The experiments were undertaken using SS301 cold-rolled to either 40% or 70% reduction in thickness at Integrity Metal and Slitting with 0.1 mm in thickness.

X-Ray diffraction studies indicate that the SS301 exposed to various magnitude of magnetic field at 77 K has significantly more volume fraction of martensite than the samples rolled to 40 %, as shown in Figure 1. Figure 2 shows the relationship between the applied fields and the volume fraction of bcc phase. The results indicate that a combination of high magnetic field and cryogenic temperatures induce a more volume fraction of ferromagnetic martensite in the samples. Therefore, SS301 indeed has the potential to transform into martensite in the magnetic field and at

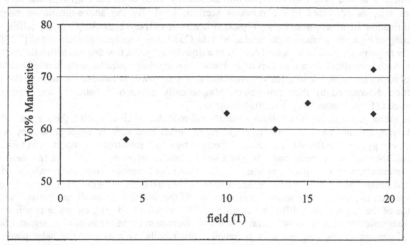

Figure 2. Relationship between the field and volume fraction of martensite of SS301 after materials were quenched to 77 K in various fields.

77K, which is the operation temperature of most of the pulsed magnets in the NHMFL. Therefore, more systematic research is undertaken to elucidate the effects systematically in more complex conditions such as a combination of high stresses and magnetic field at 77K.

Magnetic Field Induced Texture

High magnetic field can modify the texture in both bulk and thin film materials [30, 31, 32, 33, 34, 35, 36, 37, 38]. The high field studies have been done on both metallic and ceramic materials. It is gradually realized that HMFA in a high field is a useful method for engineering the texture of materials. However, the cost is an issue if only resistive magnet is used.

The earlier work was undertaken in materials with permanent magnetic properties and spinodal decompositions [39, 40, 41]. A typical example is an alnico magnet. In this system, a spinodal decomposition occurs. If the aging for decomposition is carried out just below the intersection (on the phase diagram) of the Curie temperature and the temperature where the spinodal instability takes place, the magnetic field is most effective in forming uniformly spaced rods parallel to the magnetization [40]. In order to have the best alignment and most uniformed spacing, the fields have to be adjusted according to the composition of the alloys.

In a two-phase or multi-phase material, one may expect to align one of the phases provided two phases have different magnetic properties. If the matrix phase has a very low strength and modulus, such as a liquid, the other phases can move freely and may align along the magnetic field due to its magnetocrystalline anisotropy. In this case, to align a paramagnetic or ferromagnetic particle, the drive force (magnetic torque) due to magnetocrystalline anisotropy of the new phases in a paramagnetic or ferromagnetic state could still be larger at high magnetic fields than those due to the shape anisotropy and the thermal disordering effects [42]. In these cases, a hard phase-textured microstructure has lower energy than a random structure in a magnetic field. If the applied field strength is sufficiently high, the magnetically induced crystallographic alignment of the new magnetic phases can occur even in a paramagnetic state during the HMFA and the alignment is driven by a magnetocrystalline anisotropy, such as in the $Nd_2Fe_{14}B$ and fct FePt phases etc., as addressed in the previous sections [43]. By the above-mentioned method (solidification in a magnetic field), crystal orientation of $Nd_2Fe_{14}B$ Sm_2Co_{17} crystallite and texturing of high-temperature-superconductor $YBa_2Cu_3O_7$ were successfully fabricated [44]. This usually requires that solid crystallites coexist in a liquid melt just a few degrees above the melting point. Such crystallites align their axis of the highest susceptibility parallel to the applied field and act as nuclei for solidification. Hence, preferential grain growth is induced by these nuclei in a direction determined by their orientation. Magnetically anisotropic behavior was obviously observed in these magnetically solidified samples [44].

If the matrix phase has relatively large strength and modulus, such as a solid, the magnetic field must overcome the interface energy term, the energy associated with the shape anisotropy or the elastic energy, and the thermal disordering effects. The shape anisotropy is important in this case because the rotation of the second phase in a solid phase involves deformation of the materials. The deformation may be plastic or elastic. Any elastic deformation component will introduce internal stresses and results in stress concentration at the interfaces. The energy required for such deformation is related to the strength and modulus of the matrix and the aligned phase. Such a change of the orientation is difficult but still possible, especially if the new phase is still in the ferromagnetic state during the solid-state phase transformation in the presence of a magnetic field. The alignment of the new phase is achieved more readily by changing the habit plane for nucleation of the new phase than rotating the crystal during the growth in the presence of a high magnetic field. For instance, it was found that annealing the cold-rolled Fe-Si sheet [1] in a high magnetic field of 10 T parallel to the rolling direction enhanced the selectivity of the <001> axis alignment. High magnetic field (1.5 T) annealing in another system, Armco iron, has also been studied [45]. The grains with <100> parallel to the external field would nucleate preferentially. Recently, we successfully obtained obvious enhancement of the out-of-plane (001) texture of the magnetically hard FePt phase by HMFA of cold-rolled nanolaminated Fe/Pt samples in the presence of an out-of-plane 19 T field [46,47]. The effects of HMFA (35 T) on the preferred orientation distribution (texture) and magnetic anisotropy of arc-melt FePd alloy were studied in another case. Compared with annealing without a magnetic field, the HMFA increases the volume fraction of grains with their c-axes aligned along the applied field direction [48].

Effects of Magnet Annealing on Properties

Magnetic annealing or processing materials in a magnetic field can alter the properties of the materials. Therefore, researchers make significant effort in engineering materials properties by magnetic field. For instance, microhardness measurements demonstrated that the hardness values of the SS301 exposed to 30 T magnetic fields at 77K increased from 278±21HV (in as-rolled samples) to 303±6HV. This was due to the high field induced martensite phase transformations.

76

Table I listed examples of magnetic field effects on material properties. One of the examples is the increase of hardness by 8% at the field gradient of 50 T/m [17] in pearlite due to the presence of magnetic field during the austenite – pearlite phase transformation. The following paragraphs describe examples of magnetic annealing enhancement of magnetic properties.

Compared with the melt-spun $Nd_{2.4}Pr_{5.6}Dy_1Fe_{84}Mo_1B_6$ samples annealed without a field at 690°C for 20 min, there is a noticeable improvement in $_iH_c$, $(BH)_{max}$ and $4\pi M_r$ for magnetically annealed ones. Especially, $(BH)_{max}$ at 50 K and 300 K is enhanced by 43.7 % and 35.7 %, respectively, after magnetic annealing in a field of 19 T [49]. Additionally, the kink in the demagnetization curve disappears with magnetic annealing, as shown in figure 3. In this case, the magnetic-field-induced exchange coupling enhancement between the soft and hard phases were obtained in the

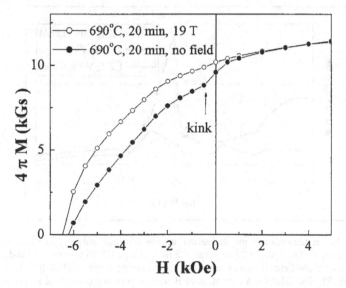

Figure 3 Room temperature demagnetization curves of two $Nd_{2.4}Pr_{5.6}Dy_1Fe_{84}Mo_1B_6$ samples annealed with or without a.19 T magnetic field at 690 °C for 20 minutes [49].

magnetically annealed sample. Such an enhancement is attributed to the nanostructure refinement. The disappearance of the kink is an indication of the magnetic-field-induced exchange coupling interaction enhancement between the soft and hard phases in the magnetically annealed sample due to the nanostructure refinement.

As the description of the kink is so important in magnetic annealed samples, we describe such an effects by plotting the magnetization curve with normalized derivative B with respect of H, as shown in Figure 4. Apparently, the significant reduction of the values of $dB/dH/ \mu_0/Br$ quantifies the disappearance of the kink and enhancement of the coupling.

The improvement of the hard magnetic properties of the magnetically annealed sample results mainly from the magnetic-field-induced crystallographic textures and the refinements of the nanostructured morphology by magnetic annealing, such as the reduced sizes and the more uniform distribution of the grains for both the hard and soft phases. Additionally, the magnetic annealing improves the crystalline quality of the hard 2:14:1 phase, which is also favorable for the magnetic hardening for the magnets.

In addition to the above-discussed $Nd_{2.4}Pr_{5.6}Dy_1Fe_{84}Mo_1B_6$ alloys, the magnetic-field-induced refinement of nanostructure and enhancement of hard magnetic properties were observed in the melt-spun nominal single-phase $Nd_2Fe_{14}B$ and $Pr_2Co_{14}B$ nanostructured magnets [50], balled-milled $SmCo_7/Co$-type nanocomposites [51], magnetron sputtered FePt films and $Nd_2Fe_{14}B/Co$ nanocompoiste thin films [52], the melt-spun $Nd_2Fe_{14}B/\alpha$-Fe-type nanocomposites with different compositions such as $Nd_{2.4}Pr_{5.6}Dy_1Fe_{84}Mo_1B_6$, $Pr_7Tb_1Fe_{87}Nb_{0.5}Zr_{0.5}B_4$, $Pr_7Tb_1Fe_{85}Nb_{0.5}Zr_{0.5}B_6$, $Nd_{2.4}Pr_{5.6}Dy_1Fe_{83.5}Mo_1Zr_{0.5}B_6$, $Nd_{2.4}Pr_{5.6}Dy_1Fe_{77}Co_6Ti_{1.5}Ga_{0.5}Zr_{0.5}B_6$ and $Nd_{2.4}Pr_{5.6}Dy_1Fe_{77.4}Co_6Mo_{1.0}Zr_{0.5}Hf_{0.1}B_6$ alloys, etc.

The texture has direct impact on materials properties. For instance, nanocomposite magnets have

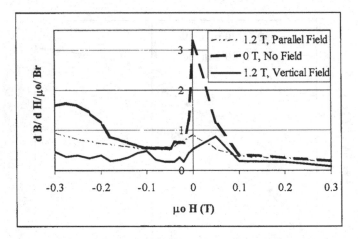

Figure 4 Room temperature magnetization curves of three $Nd_{2.4}Pr_{5.6}Dy_1Fe_{84}Mo_1B_6$ ribbons annealed at 680 °C for 1 min without a field, with a 1.2 T in-plane field and with a 1.2 T out-of-plane field. The magnetization curves were plotted by dB/dH/ μ_0/Br (1/T) verse μ_0 H (T). The dB/dH is a derivative of B with respect to applied field H. The data are normalized with μ_0Br.

extremely high theoretical energy products $(BH)_{max}$ of up to 120 MGOe [20]. However, the experimentally obtained $(BH)_{max}$ lags far behind the theoretical values [53,54,55]. Some challenges still remain in the improvement of the naonostructure morphology and crystallographic texture, such as refinement of grain sizes, homogeneity of the grain size distribution, refinement of grain configuration, and especially, the hard nanograin alignment. Homogenized and textured nanostructures are essential to make a breakthrough in hard magnetic properties of permanent magnets with nanoscale grain sizes [53,54,55]. Especially, up to now, the alignment of the hard nanograins to get strong texture and magnetic anisotropy is the core to achieve record-high $(BH)_{max}$ in both single-phase and hard/soft nanocomposite permanent magnets [53,54,55]. In other words, it is known that a significant improvement in $(BH)_{max}$ is expected from successfully fabricated anisotropic single-phase magnets or nanocomposites with crystallographic textures of hard phases [56,57]. Further research is underway towards achieving a record-high $(BH)_{max}$ by

78

preparation of homogeneously nanostructured, and especially, highly textured permanent magnets with nanoscale grain sizes.

It has been realized that the improvement of the hard magnetic properties of the magnetically annealed sample results mainly from the magnetic-field-induced crystallographic textures and the refinements of the nanostructured morphology by magnetic annealing, such as the reduced sizes of magnetic grains and the more uniform distribution of the grains for both the hard and soft phases. Additionally, the magnetic annealing improves the crystalline quality of the hard 2:14:1 phase, which is also favorable for the magnetic hardening for the magnets.

In addition to above-discussed $Nd_{2.4}Pr_{5.6}Dy_1Fe_{84}Mo_1B_6$ alloys, the magnetic-field-induced refinement of nanostructure and enhancement of hard magnetic properties were observed in the melt-spun nominal single-phase $Nd_2Fe_{14}B$ and $Pr_2Co_{14}B$ nanostructured magnets [58], balled-milled $SmCo_7$/Co-type nanocomposites [59], magnetron sputtered FePt films and $Nd_2Fe_{14}B$/Co nanocompoiste thin films [60], the melt-spun $Nd_2Fe_{14}B$/α-Fe-type nanocomposites with different compositions, treated at the same or different magnetic annealing conditions, such as $Pr_7Tb_1Fe_{87}Nb_{0.5}Zr_{0.5}B_4$, $Pr_7Tb_1Fe_{85}Nb_{0.5}Zr_{0.5}B_6$, $Nd_{2.4}Pr_{5.6}Dy_1Fe_{83.5}Mo_1Zr_{0.5}B_6$, $Nd_{2.4}Pr_{5.6}Dy_1Fe_{77}Co_6Ti_{1.5}Ga_{0.5}Zr_{0.5}B_6$ and $Nd_{2.4}Pr_{5.6}Dy_1Fe_{77.4}Co_6Mo_{1.0}Zr_{0.5}Hf_{0.1}B_6$ alloys, etc.

Table I. Selected Examples of Material Phase Transformation and Properties Influenced by High Magnetic Fields

Materials	Microstructure/Phase Transformation	Field	Properties	Reference
Fe-Pt	Texture/microstructure refinement	19 T	$(BH)_{max}$ improvement	46, 47
Fe-0.1wtC / Fe-0.6wt%C	Alignment of Paramagnetic Austenite Grains	8T		61 62
Fe-1C-Cr&/Mn	Supersaturated C in ferrite formed during the austenite to pearlite phase transition	50T/m	Hardness improvement by 8%	17
$Nd_{2.4}Pr_{5.6}Dy_1Fe_{84}Mo_1B_6$	Alignment of the hard magnetic phase	19T	$(BH)_{max}$ improvement by 43%	49

Conclusions

The effects and relevant mechanism of magnetic field annealing on nanostructures, crystallographic textures, and properties for selected materials have been discussed. Magnetic annealing promotes the phase transformation and increases the nucleation site numbers for the formation of magnetic phases. The high magnetic field heat treatment may result in the refinement of the microstructure, such as finer, less angular and more homogeneously distributed grains with higher crystalline quality, etc. Texture introduced by the high magnetic fields is equivalent important to refinement of the microstructure. Particularly, magnetic annealing in a high field is a promising method for manufacturing textured nanomagnetic materials. The improvement of properties of the magnetically annealed sample usually results from the magnetic-field-induced crystallographic textures and nanostructure refinement.

References

1 N. Masahashi, M. Matsuo, and K. Watanabe, J. Mater. Res. **13**, 457 (1998)
2 Sadovskiy V.D., Rodigin, L, Smirnov, L.V., Filonchik, G.M., Fakidov, I.G., Phys Met.,

Metallogra, 12, 131, (1961)

[3] Satyannarayan K.R., Eliasz W., Miodownik, A.P., Acta Metall, 16, 877, (1968)

[4] R. Fields C.D. Graham, "Effect of high magnetic fields on martensite transformation," *Metall. Trans. A* **7A** 719 (1976)

[5] T. Kakeshita, K. Kuroiwa, K. Shimizu, T. Ikeda, A. Yamagishi and M. Date, "Effect of magnetic fields on athermal and isothermal martensitic transformations in Fe-Ni_Mn alloys," *Mat. Trans. Jap. Inst. Met.* **34**(5) 415 (1993)

[6] P.A. Algarabel, C. Magen, L. Morellon, M.R. Ibarra, F. Albertini, N. Magnani, A. Paoluzi, L. Pareti, M. Pasquale, S. Besseghini, "Magnetic-field-induced Strain in Ni_2MnGa Melt-Spun Ribbons," *Journal of Magnetism and Magnetic Materials* **272-276** 2047-2048 (2004)

[7] M. Pasquale, C.P. Sasso, S. Besseghini, E. Villa and V. Chernenko, "Temperature Dependence of Magnetically Induced Strain in Single Crystal Samples of Ni-Mn-Ga," *Journal of Applied Physics* **91**(10) 7815-7817 (2002)

[8] J. Xunyong, X. Huibin, Gao Xueping and S. Deying, "Influence of Magnetic Field on the Phase Transformation of TbDyFe/NiTi Composite Film," *Journal of Materials Science Letters* **22** 729-731 (2003)

[9] F. Chen, Z.Y. Gao, W. Cai, L.C. Zhao, G.H. Wu, J.L. Chen and W.S. Zhan, "Strains Induced by Magnetic Field and Phase Transformationin $Ni_{50.5}Mn_{26.2}Ga_{23.4}$ Ferromagnetic Shape Memory Alloy," *Journal of Materials Science Letters* **22** 1241-1242 (2003)

[10] N. Nersessian, S.W. Or, G.P. Carman, S.K. McCall, W.C hoe, H.B. Radousky, M.W. McElfresh, V.K. Pecharsky and A.O. Pecharsky, "$Gd_5Si_2Ge_2$ Composite for Magnetostrictive Actuator Applications," *Applied Physics Letters* **84**(23) 4801-4803 (2004)

[11] V.C. Solomon, D.J. Smith, Y. Tang, A.E. Berkowitz, "Microstructural Characterization of Ni-Mn-Ga Ferromagnetic Shape Memory Alloy Powders," *Journal of Applied Physics* **95**(11) 6954-6956 (2004)

[12] K. Inoue, K. Enami, M. Igawa, Y. Yamaguchi, K. Ohoyama, Y. Morii, Y. Matsuoka and K. Inoue, "Possibility of Controlling the Shape Memory Effect by Magnetic Field," *International Journal of Applied Electromagnetic and Mechanics* **12** 25-33 (2000).

[13] N. Lansky, O.Söderberg, A. Sozinov, Y. Ge, K. Ullakko and V.K. Lindroos, "Composition and Temperature Dependence of the Crystal Structure of Ni-Mn-Ga Alloys," *Journal of Applied Physics* **95**(12) 8074- 8078 (2004).

[14] S.-Y. Chu, A. Cramb, M. De Graef, D. Laughlin and M.E. McHenry, "The Effect of Field Cooling and Field Orientation on the Martensitic Phase Transformation in a Ni_2MnGa Single Crystal," *Journal of Applied Physics* **87**(9) 5777-5779 (2000).

[15] T.W. Shield, "Magnetomechanical Testing Machine for Ferromagnetic Shape-Memory Alloys," *Review of Scientific Instruments* **74**(9) 4077-4088 (2003).

[16] V.N. Pustovoit, Yu. M. Dombrovskii and S.A. Grishin. *Metalloved Term Obrab Met*, p. 22, (1979).

[17] Shimotomai M, Influence of magnetic-field gradients on the pearlitic transformation in steels MATERIALS TRANSACTIONS 44 (12): 2524-2528 DEC (2003)

[18] M. Enomoto, H. Guo, Y. Tazuke, Y.R. Abe and M. Shimotomai. *Metall Mater Trans* **32A**, 445 (2001)

[19] C.C.Koch, Materials Science and Engineering, A287, 213-218, (2000).

[20] Gao MC,Bennett TA , Rollett AD,Laughlin DE , The effects of applied magnetic fields on the alpha/gamma phase boundary in the Fe-Si system, JOURNAL OF PHYSICS D-APPLIED PHYSICS 39 (14): 2890-2896 JUL 21 2006.

[21] Christern, Phase Transformation, (1995).

[22] N.Saunders N, Miodonik AP, Calphas, Oxford, Elsevier Science, 253-8, (1998)

[23] T. Kakeshita, T. Saburi, and K. Shimizu, Mater. Sci. & Eng. A **273-275**, 21 (1996).

[24] H. Ohtsuka, X. J. Hao, and H. Wada, International Workshop on Materials Analysis and

Processing in Magnetic Fields, Tallahassee, FL, March 2004.

[25] T. M. Zhao, Y. Y. Hao, X. R. Xu, Y. S. Yang and Z. Q. Hu, J. Appl. Phys. **85**, 518 (1999)
[26] Dong J, Li ZF, Zeng XQ, Lu C and Ding WJ, MAGNESIUM - SCIENCE, TECHNOLOGY AND APPLICATIONS MATERIALS SCIENCE FORUM 488-489: 849-852 2005
[27] H. Y. Wang, X. K. Ma, Y. J. He, S. Mitani, and M. Motokawa, Appl. Phys. Lett **85**, 2304 (2004).
[28] T. Klemmer, D. Hoydick, H. Okumura, B. Zhang, and W. A. Soffa, Scripta Metall. Mater. **33**, 1793 (1995).
[29] H. Y. Wang, X. K. Ma, Y. J. He, S. Mitani, and M. Motokawa, Appl. Phys. Lett **85**, 2304 (2004)
[30] F. Gaucherand, and E. Beaugnon, "Magnetic Texturing in Ferromagnetic Cobalt Alloys," Physica B **346-347** 262-266 (2004)
[31] J.E. Evetts, "Transport Properties of Low-Angle Grain Boundaries and Granular Coated Conductors," Superconductor Science and Technology 17 S315-318 (2004)
[32] S. Awaji, Y. Ma, W.P. Chen, H. Maeda, K. Watanabe and M. Motokawa, "Magnetic Field Effects on Synthesis Process of High-T_c Superconductors," *Current Applied Physics* **3** 391-395(2003)
[33] A. Makiya, D. Kusano, S. Tanaka, N. Uchida, K.; Uematsu, T. Kimura, K. Kitazawa and Y. Doshida, "Particle Oriented Bismuth Titanate Ceramics Made in High Magnetic Field," *Journal of the Ceramic Society of Japan* **111**(9) 702-704 (2003)
[34] L. Wu, S.V. Solovyov, H.J. Wiesmann, D.O. Welch and M. Suenaga, "Twin Boundaries and Critical Current Densities of $YBa_2Cu_3O_7$ Thick Films Fabricated by the BaF_2 Process," *Superconductor Science and Technology* **16** 1127-1133 (2003)
[35] V. Selvamanickam, A. Goyal and D.M. Kroeger, "Grain Alignment in Bulk $YBa_2Cu_3C_x$ Superconductor by a Low Temperature Phase Transformation Method," *Applied Physics Letter* **65**(5) 639-641 (1994)
[36] T. Uchikoshi, T.S. Suzuki, H. Okuyama and Y. Sakka, "Control of Crystalline Texture in Polycrystalline Alumina Ceramics by Electrophoretic Deposition in a Strong Magnetic Field," *Journal of Materials Research* **19**(5) 1487-1491 (2004)
[37] P.D. Rango, M. Lees, P. Lejay, A. Sulpice, R. Tournier, M. Ingold, P. Germi and M. Pernet, "Texturing of Magnetic Materials at High Temperature by Solidification in a Magnetic Field," *Nature* **349**(6312) 770-771 (1991)
[38] H. Maeda, P.V.P.S.S. Sastry, U.P. Trociewitz, J. Schwartz, K. Ohya, M. Sato, W.P. Chen, K. Watanabe and M. Kmotokawa, "Effect of Magnetic Field Strength in Melt-Processing on Texture Development and Critical Current Density of Bi-oxide Superconductors," *Physica C* **386** 115-121 (2003)
[39] Jones B, von Emden HJM. Philips Tech Rev;6:8. (1941)
[40] Cahn J.W., J Appl Phys ;34:3581 (1963).
[41] Iwama Y, Takeuchi M. Trans Jpn Inst Met;15:371 (1974)
[42] B. A. Legrand, D. Chateiger, R. P. de la Bâthie, and R. Tournier, J. Magn. Magn. Mater. **173**, 20 (1997).
[43] P. Courtois, R. P. de la Bâthie, and R. Tournier, J. Magn. Magn. Mater. **153**, 224 (1996).
[44] P. de Rango, M. Lees, P. Lejay, A. Sulpice, R. Tournier, M. Ingold, P. Germ, and M. Pernet, Nature **349**, 770 (1991)
[45] H. O. Martikainen, and V. K. Lindroos, Scandinavian Journal of Metallurgy **10**, 3 (1981).
[46] B.Z.Cui; K. Han, Garmestani, H.; Schneider-Muntau, H.J., "Structure and magnetic properties of FePt and FePt–Ag nanostructured magnets by cyclic cold rolling", J. Applied Physics, **99** (8), 08-910 (2006).
[47] B.Z.Cui; K. Han, Li, D.S.; Garmestani, H.; Liu, J.P.; Dempsey, N.M. and Schneider-Muntau, H.J., *Magnetic-field-induced Crystallographic Texture Enhancement in Cold-deformed FePt*

Nanostructured Magnets, J. Applied Physics, **100** (1), 013902 (2006).

[48] D. S. Li, H. Garmestani, Shi-shen Yan, M. Elkawni, M. B. Bacaltchuk, H. J. Schneider-Muntau, J. P. Liu, S. Saha and J. A. Barnard, J. Magn. Magn. Mater. **281**, 272 (2004).

[49] B.Z.Cui; K. Han, Garmestani, H.; Su, J.H.; Schneider-Muntau, H.J. and Liu, J.P., Enhancement of Exchange Coupling and Hard Magnetic Properties in Nanocomposites by Magnetic Annealing, Acta Mater., 53, 4155-4161 (2005)

[50] Y. Otani, H. Sun, J. M. D. Coey, H. A. Davis, A. Manaf, and R. A. Buckley, J. Appl. Phys. **67**, 4616 (1990).

[51] B. Z. Cui, M. Q. Huang, S. Liu, IEEE. Trans. Magn. **39** 2866 (2003).

[52] M. J. O'Shea and B. Z. Cui, IEEE. Trans. Magn. **40** 2889 (2004).

[53] M. Jurczyk, S. J. Collocott, J. B. Dunlop and P. B. Gwan, J. Phys. D: Appl. Phys. **29**, 2284 (1996).

[54] B. Z. Cui, X. K. Sun, L Y. Xiong, W. Liu, Z. D. Zhang, Z. Q. Yang, A. M. Wang and J. N. Deng, J. Mater. Res. **16**, 709 (2001).

[55] A. Manaf, R. A. Buckley, H. A. Davies, J. Magn. Magn. Mater. **128**, 302 (1993).

[56] R. Skomski and J. M. D. Coey, Phys. Rev. B **48**, 15812 (1993).

[57] R. Fischer, T. Schrefl, H. Kronmüller and J. Fiddler, J. Magn. Magn. Mater. **150**, 329 (1995).

[58] Y. Otani, H. Sun, J. M. D. Coey, H. A. Davis, A. Manaf, and R. A. Buckley, J. Appl. Phys. **67**, 4616 (1990).

[59] B. Z. Cui, M. Q. Huang, S. Liu, IEEE. Trans. Magn. **39** 2866 (2003).

[60] M. J. O'Shea and B. Z. Cui, IEEE. Trans. Magn. **40** 2889 (2004).

[61] Maruta K, Shimotomai M. Mater Trans JIM ;41:902 (2000).

[62] Shimotomai M, Maruta K. Scr Mater;42:499 (2000).

Materials Processing under the Influence of External Fields
Edited by Qingyou Han, Gerard Ludtka, Qijie Zhai
TMS (The Minerals, Metals & Materials Society), 2007

SYNTHESIS OF ACICULAR MAGNETITE USING COPRECIPITATION

METHOD UNDER MAGNETIC FIELD

Xi Yun Yang, Bai Zhen Chen,Hui Xu

School of Metallurgical Science and Engineering, Central South University, Changsha 410083, China

Keywords: Acicular magnetite, Magnetic field, Precipitation

Abstract

Acicular magnetite was synthesized by precipitation of ferrous and ferric mixed salt solution with urea in the temperature range 90-95°C in reflux condition for 12h. The magnetite particles were characterized by X-ray powder diffraction, XPS, scanning electron microscopy and magnetization measurements. XRD showed that the mole ratio of Fe^{3+}/Fe^{2+} had great effect on the composition of the product. SEM showed that the type of anion didn't obviously affect the shape of magnetite particles,magnetic field played a crucial role to form acicular magnetite. Acicular magnetite with diameter of 100nm and length of about 0.5μm was formed under a 0.3T magnetic field The hysteresis loop of magnetite showed a ferromagnetic behavior with saturation magnetization of 75.6emu/g and coercivity of 380 Oe.

Introduction

Acicular magnetite(Fe_3O_4) particles possess unique magnetic and electronic properties, they are very widely used for recording material[1],ferrofluids[2] and electrophotographic developer[3],etc. There are many methods reported for the synthesis of acicular Fe_3O_4 particles, e.g. reduction of hematite by CO/CO_2[4] or H_2[5],hydrothermal synthesis using surfactant as template[6].Recently, magnetic field assisted growth of one-dimensional magnetic nanostructured materials has been the subject of research[7].Single crystalline Fe_3O_4 nanowire could be formed by a hydrothermal process under the induction of an applied external field[8]. However, in all these reported methods, the experiment conditions need to be carefully controlled, otherwise mixture of oxides is obtained, In this paper, we report preparation of acicular magnetite by a homogeneous precipitation under an external magnetic field, which did not require surfactant and rigorous conditions.

Experimental

The chemical reagents used in the study were ferrous sulfate, urea and ferric sulfate . All the chemicals were analytical grade. Ferrous sulfate and ferric sulfate were dissolved in distilled water and adjusted to a total iron concentration of 0.20mol/L. A urea solution with a concentration of 1.0mol/L was also prepared. For each experiment, an appropriate volume of urea solution was added into ferrous and ferric mixed solution, during the process, nitrogen was kept passing through the solution to prevent the oxidation of Fe^{2+} in the system. After being blown in for 20min,the suspension was transferred to a flask with reflux condenser and heated at 90□ for 12h, and then cooled to room temperature naturally. After that, the resulting black precipitate was filtered, washed with distilled water three times and finally dried in air at 80□.

The prepared particles were examined by the following conventional techniques. Powder X-ray diffraction patterns were recorded on a Rigaku diffractometer using CuKαradiation. X-ray photoelectron spectroscopy(XPS) was carried out using a Kratos XSAM800 spectrometer using MgKαradiation, the binding energies of sample were calibrated using C1s peak as reference(284.8ev). Morphology of the particles was observed with a JEOL 6060LV scanning electron microscope. Magnetic properties were measured by physical property measurement system (PPMS) under a magnetic field of 5000Oe. The ferrous and total iron contents were determined by titrating with $K_2Cr_2O_7$, thus the Fe^{3+}/Fe^{2+} ratio of sample was obtained.

Results and discussion

Preparation of acicular magnetite

The acicular magnetite has been synthesized by hydrolysis of $Fe^{2+}$and Fe^{3+} solution in the presence of urea at 90 °C for 12h.The initial molar ratio of Fe^{3+} to Fe^{2+} has great effect on phase compositions of the product. Fig.1 is the XRD patterns of the product obtained at different Fe^{3+} /Fe^{2+} ratio. When the ratio is 1:1,the product was a mixture of Fe_2O_3,FeOOH,Fe_3O_4,the appearance of Fe_2O_3 and FeOOH was due to hydrolysis of excessive Fe^{3+} ions. For ratio of 1:2,the product was pure magnetite without any impurity phases.

It has been reported that anion may absorb on some crystal faces of oxide, thus affects the growth of crystal in some direction and obtains oxide with different shape. In the experiment, some type of anion was added into the suspension. Fig.3 shows the effects of anion on shape of magnetite. In the presence of ferrous sulfate or ferrous chloride, the particles were all spherical, but in a solution containing ferrous sulfate and KNO_3, the particles were spherical and square. So anion can't play great role to obtain acicular magnetite. Therefore, it was difficult to obtain acicular magnetite directly from aqueous solution.

84

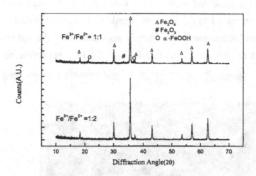

Fig.1 X-ray diffraction spectra of the sample prepared at different Fe^{3+} /Fe^{2+} ratio

Figure 2. Effect of anion type on the shape of sample

It is well known that magnetic material possess easy magnetic axes, Fe_3O_4 ,for example, possess easy magnetic axes.Fe_3O_4,for example, has easy magnetic axes along the [111] and [110] directions. Therefore, an applied external magnetic field might induce oriented growth, for example along the easy magnetic axis, which could result in the formation of 1D structures of magnetic materials.Fig.4 shows SEM images of the samples prepared under external magnetic field, which clearly revealed the effect of magnetic field on the growth process of Fe_3O_4.Without a magnetic field applied, Fe_3O_4 particles were spherical in shape(Fig.2), the morphology changed drastically when an external magnetic field was applied. In addition to few spherical particles, acicular-like particles were observed in the sample when a 0.1T magnetic field was applied. When the strength was increased to 0.3T,a large number of acicular particles formed (diameter 100nm and length 0.5μm),accompanied by the disappearance of the spherical particles. The results clearly show that the external magnetic field could be responsible for the formation of acicular particles.

Figure 3. SEM image of the samples obtained in 0.1T(A) and 0.3T(B) magnetic field

Structure

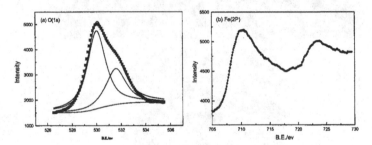

Figure 4. XPS spectra of (a) O(1s) and (b)Fe(2p) of sample

In order to have a deeper insight into the surface characterization of particles, XPS analysis was carried out, which provides information on the chemistry of the constituents. The binding energy(BE) of O(1s) line for magnetite is 530.0ev,whereas for hydroxide, in general, it is higher by 1.5-2.5ev in comparison to the corresponding oxides. The O(1s) spectrum had two components(Fig.4a). The peak at 529.9ev is ascribed to Fe_3O_4, the second peak at 531.3ev is attributed to the hydroxyl species from the chemisorbed water and CO_2, because C1s(its binding energy is 284.8ev) was taken as a standard in our experiments. The XPS spectrum of Fe2p is shown in Fig.4b.The spectrum was split by spin-orbit interactions into $Fe2p_{3/2}$ and $Fe2p_{1/2}$ peaks. The corresponding BE were located at 710.4 and 723.8ev, respectively. It is well known that Fe_3O_4 particles are chemically unstable and easily oxidized to Fe_2O_3 in air. The oxides of Fe for Fe_2O_3 and FeO display shake up satellites at 720 and 716ev, respectively, while Fe_3O_4 does not display any satellite[9]. No satellite was observed between $Fe2p_{3/2}$ and $Fe2p_{1/2}$ peaks for the sample shown in Fig.4b. The results suggest that the particles exhibit the high resistance to air-oxidation.

Magnetic properties measurement

Fig.5 shows magnetic hysteresis curves for the samples measured at room temperature. It shows that sample A and B had saturation magnetizaton of 76.2 emu/g and 75.7 emu/g, respectively, coercivity of 270 Oe and 380Oe, respectively. Due to the large shape anisotropy, the coercity was high, the sample could be suited for magnetic recording material.

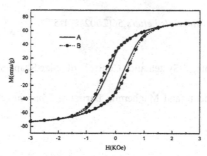

Figure 5. Magnetic hysteresis curves of sample prepared in 0.1T(A) and 0.3T(B) magnetic field

Conclusion

Acicular magnetite was synthesized by precipitation of ferrous and ferric mixed salt solution with urea in the temperature range 90-95°C in reflux condition for 12h. The mole ratio of Fe^{3+}/Fe^{2+} had great effects on the composition of the product. By maintaining mole ratio of Fe^{3+}/Fe^{2+} of 1:2, stoichiometric Fe_3O_4 can be obtained. The type of anion didn't obviously affect the shape of magnetite particles, external magnetic field played a crucial role to form acicular magnetite. Acicular magnetite with diameter of 100nm and length of about 0.5μm was formed under a 0.3T magnetic field. Magnetite particles had a saturation magnetization of 75.7emu/g and coercivity of 380 Oe, which were expected to be used for magnetic recording material.

References

[1] C.P.Hancock, et al., "Characterization of Magnetic Pigment Dispersions Using Pulsed Magnetic Fields, "IEEE Transactions on Magnetic,32(1996)4037-4039.

[2] T.Nakamura, H.Sadamura, " Surface Modification of γ-Fe_2O_3 Particles with Cobalt

Containing Spinel, "Physica Status Solidi(A) Applied Research,155(1996)207-213.

[3] U. Meisen and H.Kathrein, "The Influence of Particle Size,Shape and Particle Size Distribution on Properties of Magnetite for the Production of Toners, "*Journal of Imaging Science and Technology*, 44(2000)508-513.

[4] Y.B.Khollam, et al.,"Microwave Hydrothermal Preparation of Submicron-sized Spherical Magnetite(Fe$_3$O$_4$) Powders," *Materials Letters*,56(2002)571-577.

[5] H. Itoh and T.Sugimoto, "Systematic Control of Size,Shape,Structure,and Magnetic Properties of Uniform Magnetite and Maghemite Particles, "*Journal of Colloid and Interface Science* , 265(2003)283-295.

[6] J.Wan,et al., "A Soft-template-Assisted Hydrothermal Approach to Single-Crystal Fe$_3$O$_4$ Nanorods, "*Journal of Crystal Growth*,276(2005)571-576.

[7] J.Wang,et al., "Growth of Magnetite Nanorods along its Easy-Magnetization Axis of [110], " *Journal of Crystal Growth*,263(2004)616-619.

[8] J.Wang,et al., "Magnetic-Field-Induced Growth of Single-Crystalline Fe$_3$O$_4$ Nanowires, " *Advanced Materials*,16(2004)137-139.

[9] C.Ruby,J.Fusy and J.-M.R.Genin, "Preparation and Characterisation of Iron Oxide Films Deposited on MgO(100), "*Thin Solid Films*,352(1999) 22-28.

Materials Processing under the Influence of External Fields
Edited by Qingyou Han, Gerard Ludtka, Qijie Zhai
TMS (The Minerals, Metals & Materials Society), 2007

RAPID CERAMIC PROCESSING BY FIELD-ASSISTED SINTERING TECHNIQUE

Dat V. Quach[1], Veaceslav Zestrea[2], Vladimir Y. Kodash[1], Gregory Toguyeni[3], and Joanna R. Groza[1]

[1]Department of Chemical Engineering and Materials Science, University of California, Davis
One Shields Avenue, Davis, CA 95616, USA
[2]Institute of Applied Physics, Academy of Science of Moldova
MD-2028 Chisinau, Republic of Moldova
[3]Departement Science des Materiaux, Ecole Polytechnique de l'Universite de Nantes, France
rue Christian Pauc – BP 50609, Nantes Cedex 3 - France

Keywords: reaction sintering, field-assisted sintering, magnetic materials

Abstract

The novel field-assisted sintering technique (FAST) has been successfully used to synthesize and consolidate a number of inorganic compounds such as chalcogenide spinels and tialite. A ternary magnetic sulfide compound $FeCr_2S_4$ was prepared from a mixture of commercially available FeS and Cr_2S_3 powders using FAST. A one-step complete reaction sintering to form $FeCr_2S_4$ was done at 1000°C in 10 minutes. The ternary magnetic oxide compound $CoAl_2O_4$ was also prepared using the same technique. The high homogeneity of the composition of the prepared samples was confirmed by XRD analysis. In addition, the hard-to-form-and-sinter tialite, Al_2TiO_5, can also be FAST-synthesized from regular Al_2O_3 and TiO_2 powders. Compared to conventional methods where the same reaction sintering step may take hours or days, current FAST-reaction sintering experiments demonstrate the ability of this novel technique to perform fast reaction sintering of different ceramic compounds.

Introduction

Field-assisted sintering technique (FAST) is a non-conventional powder consolidation method in which an external electric field / electrical current is combined with resistance heating and modest pressure (up to 100 MPa) to enhance densification. Most commonly a high DC pulsed current is used as an external energy source. This technique employs a very high heating rate (50 to 1000°C min^{-1}) that significantly reduces the high-temperature exposure and restrains grain growth. A detailed description of FAST is provided elsewhere [1]. It has been assumed that either sparks or plasma are generated and can lead to the removal of surface impurities during the initial stage of sintering [2-4]. This is a possible reason for enhanced sintering behavior associated with FAST that has been widely reported. Although the concept of spark/plasma formation is plausible, the existence of such phenomenon has not been experimentally confirmed, and the role of pulsed current remains a research challenge.

There are many ceramic materials produced by reaction sintering. In this process, the initial powders comprise a mixture of powders of different phases or of powders that undergo a reaction upon thermal exposure. This is a combined one-step process (reaction + sintering), and the calcination step may be skipped. While the processing time may be reduced significantly by this combination, the major issue is related to possible incomplete reaction and inhomogeneous structure of the products. The application of an electric field has been shown to influence crystal defect motion, crystal nucleation and growth, evaporation and oxidation processes [5-9]. In addition, high current density could also promote defect formation and enhance mass transport [10-12]. FAST processing has been successfully utilized to produce highly homogeneous ceramics (SiAlON [13]) and intermetallic compounds (Nb$_3$Al [14]). The purpose of this investigation is to study the ability of FAST to synthesize other ceramics including magnetic semiconducting FeCr$_2$S$_4$, ternary magnetic oxide CoAl$_2$O$_4$ and structural Al$_2$TiO$_5$ via the reaction sintering route in a cost and time efficient manner.

Experiments

Most powders (FeS, Cr$_2$S$_3$, Co$_3$O$_4$ and TiO$_2$) were obtained from Alfa Aesar (Ward Hill, Massachusetts, USA). The Al$_2$O$_3$ powder used in (Co$_3$O$_4$ + Al$_2$O$_3$) experiments was also from Alfa Aesar while the Sumitomo AKP-500 Al$_2$O$_3$ powder (Tokyo, Japan) was used to form Al$_2$TiO$_5$. All powders are either submicron (AKP-500 Al$_2$O$_3$) or micron (the rest of the powders) in size. The stoichiometric mixtures of the powders were prepared for the following reactions:

$$FeS_{(s)} + Cr_2S_{3(s)} = FeCr_2S_{4(s)} \tag{1}$$
$$2Co_3O_{4(s)} + 6Al_2O_{3(s)} = 6CoAl_2O_{4(s)} + O_{2(g)} \tag{2}$$
$$Al_2O_{3(s)} + TiO_{2(s)} = Al_2TiO_{5(s)} \tag{3}$$

The powders were mixed in a plastic bottle with ZrO$_2$ balls for three hours. After mixing, about 3 g of each powder mixture was placed inside a graphite die with an inner diameter of 19 mm and was ready for reaction sintering. All sintering experiments were performed in "Dr. Sinter" (Sumitomo, Model 1050, Coal Mining Company, Japan). The chamber pressure was maintained at about 10 Pa, and all temperatures were measured on the die surface by a pyrometer. The pulsing sequence was 12-on and 2-off, with each pulse duration of about 3.3 ms. The sintering parameters for each experiment are listed on Table I. Phase identification of the sintered products was done using X-ray diffraction (XRD) method (Scintag XDS 2000 diffractometer with CuKα as an X-ray source). Scanning electron micrographs of the fracture surface of the Al$_2$TiO$_5$ sample were obtained using a FEI XL30 SFEG SEM (Hillsboro, Oregon, USA).

Table I. FAST Reaction Sintering Parameters

Sample	Initial Powder Mixture	Temperature	Holding Time	Mechanical Pressure
I	FeS + Cr$_2$S$_3$	1000°C	10 min.	57 MPa
II	2Co$_3$O$_4$ + 6Al$_2$O$_3$	1000°C	15 min.	57 MPa
III	2Co$_3$O$_4$ + 6Al$_2$O$_3$	1000°C	20 min.	32 MPa
IV	Al$_2$O$_3$ + TiO$_2$	1100°C	10 min.	57 MPa
V	Al$_2$O$_3$ + TiO$_2$	1200°C	10 min.	57 MPa
VI	Al$_2$O$_3$ + TiO$_2$	1300°C	10 min.	57 MPa

Results and Discussion

Figure 1 shows the XRD patterns of sample I. While the initial powder mixture consisted of FeS and Cr$_2$S$_3$, the product contained only FeCr$_2$S$_4$. The sintered specimen was homogeneous, and no residual FeS and Cr$_2$S$_3$ could be found.

Presently, single crystals of FeCr$_2$S$_4$ or chalcogenide spinels in general can be grown by either high temperature solution or chemical vapor transport. The former method involves crystallization of the spinels from metallic or salt melts. While this process is feasible to produce chalcogenide spinels with a few mm in size, contamination with solvents is usually an issue [15]. Chemical vapor transport growth, on the other hand, requires longer time. A typical growth of one single crystal of this type of materials can take from 14 to 30 days [15, 16]. Bulk polycrystalline FeCr$_2$S$_4$ can be produced in two steps. The first step often takes hours or days to prepare the spinel powder either by high-temperature solid state synthesis from elemental mixtures/binary compounds or by sol-gel method [17, 18]. The second step is the sintering of the spinel powder in which the decomposition of $FeCr_2S_4$ may occur due to high pressure and long exposure time in conventional sintering methods. Above 1.5 GPa and 830°C, $FeCr_2S_4$ undergoes a polymorphic transformation into a new phase having the structure of either NiAs or Cr_3S_4 [15]. FAST sintering indicated complete reaction of FeS and Cr_2S_3 powders to transform into $FeCr_2S_4$ in only 10 minutes. Even though the sintering temperature in FAST was quite high, no polymorphic transformation of $FeCr_2S_4$ has occurred. This can be explained by the fact that our FAST experiments employed a moderate pressure (57 MPa compared to a few GPa in hot pressing); therefore, the spinel phase was preserved because it has a lower density compared to other polymorphic phases. The short exposure time at high temperature in FAST may also avoid the decomposition of $FeCr_2S_4$.

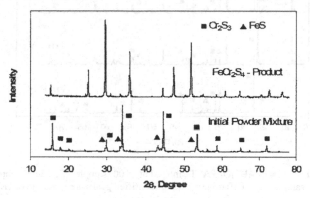

Figure 1. XRD patterns of $FeCr_2S_4$ FAST-sintered at 1000°C for 10 minutes.

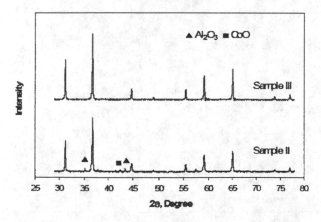

Figure 2. XRD patterns of FAST-sintered $CoAl_2O_4$ (the remaining unidentified peaks are those of $CoAl_2O_4$).

XRD patterns of samples II and III (Figure 2) demonstrate that bulk $CoAl_2O_4$ can also be formed and consolidated by FAST in less than 30 minutes. Sample II was sintered at 1000°C for

91

15 minutes, and a small amount of Al_2O_3 and CoO was detected by XRD. As the sintering time was increased to 20 minutes and the mechanical pressure was reduced from 57 to 32 MPa, a complete reaction between Co_3O_4 and Al_2O_3 was achieved (in sample III). During the sintering, Co_3O_4, unstable at high temperature, decomposed into CoO and O_2. While a longer sintering time allows a complete reaction between CoO and Al_2O_3, a lower mechanical pressure may allow O_2 gas to escape and speed up the decomposition of Co_3O_4.

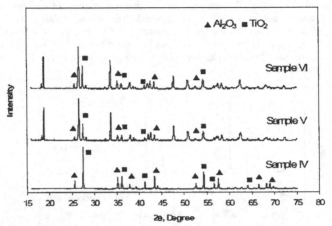

Figure 3. XRD patterns of Al_2TiO_5 FAST-sintered at 1100 (sample IV), 1200 (sample V) and 1300°C (sample VI) for 10 minutes; all unidentified peaks are those of Al_2TiO_5.

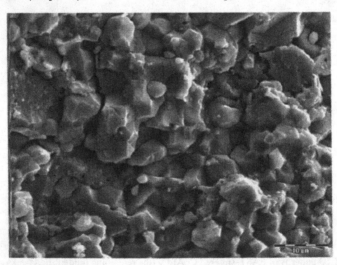

Figure 4. SEM image of the fracture surface of Al_2TiO_5 FAST-sintered at 1200°C for 10 min.

Figure 3 shows diffraction patterns of samples IV, V and VI. Only Al_2O_3 and TiO_2 were present in sample IV, and no reaction has occurred in this sample. In contrast Al_2TiO_5 was found in samples V and VI. However, there was no significant difference in terms of the Al_2TiO_5 content between these two samples even though the sintering temperature was increased from 1200°C in sample V to 1300°C in sample VI. The reaction yield in the last two experiments was evaluated based upon XRD data and was 70-75%. The fracture surface of the FAST-sintered specimen (Figure 4) shows a highly dense material with microcracks which are very typical for Al_2TiO_5.

Further investigations are needed before we are able to fully understand the reaction sintering behavior of Al_2TiO_5 and reach an optimal sintering condition for this material. In our previous study, higher reaction yield (92.5%) of Al_2TiO_5 was achieved in FAST with a special nano Al_2O_3-TiO_2 powder mixture prepared by sol-gel method [19]. However, even in that study a similar trend was noticed that an increase in temperature beyond a critical value does not give rise to an increase in reaction yield. In addition, conventional heating of the same nano Al_2O_3-TiO_2 sol-gel powder mixture at the same temperature (but for a longer time, 2 hours) yielded only 70 wt% of Al_2TiO_5 [19]. The results indicate that FAST has enhanced the reaction between Al_2O_3 and TiO_2.

Conclusions

For the first time, bulk polycrystalline $FeCr_2S_4$ was prepared by FAST in only 10 minutes from commercial FeS and Cr_2S_3 powders. The sintered product was homogeneous, and no decomposition of $FeCr_2S_4$ has occurred. FAST is also capable of producing $CoAl_2O_4$ and Al_2TiO_5 from constituent binary oxides even though further studies are essential before we can obtain an optimal reaction sintering for Al_2TiO_5. The current experiments demonstrate the strong capability of FAST in reaction sintering. Ceramic materials with complicated composition can be obtained and fully consolidated from commercially available constituent powders in a shorter time.

References

1.	J.R. Groza. "Field Assisted Sintering," *ASM Handbook*, vol. 7 (Metals Park, OH: The Materials Information Society, 1998), 583-589.
2.	M. Tokita, "Development of Large-Size Ceramic/Metal Bulk FGM Fabricated by Spark Plasma Sintering." *Mater. Sci. Forum*, 308-311 (1999), 83-89.
3.	S.H. Risbud, and J.R. Groza, "Clean Grain Boundaries in Aluminum Nitride Ceramics Densified without Additives by a Plasma-Activated Sintering Process," *Philos. Mag. B*, 69 (3) (1994), 525-533.
4.	J.R. Groza, M. Garcia, and J.A. Schneider, "Surface Effects in Field-Assisted Sintering." *J. Mater. Res.*, 16 (1) (2001), 286-292.
5.	Z.A. Munir and H. Schmalzried. "The Effect of External Fields on Mass-Transport and Defect-Related Phenomena," *J. Mater. Synth. Process.*, 1 (1) (1993), 3-16.
6.	Y.K. Min, Z.A. Munir, and H. Schmalzried, "Influence of an Electric Field on the Relative Abundance of Dislocation Lamellae in Sodium Chloride," *Philos. Mag. A*, 71 (4) (1995), 815-29.
7.	D. Kashchiev, "Nucleation in External Electric Field," *J. Cryst. Growth*, 13/14 (1972), 128-130.
8.	Z.A. Munir, and A.A. Yeh, "Evaporation of KCl Crystals in the Presence of AC and DC Fields," *Philos. Mag. A*, 56 (1) (1987), 63-71.
9.	C.A. Machida, and Z.A. Munir, "The Development of Thermal Etch Pits on Cleaved NaCl Crystal in the Presence of and Electric Field." *J. Cryst. Growth*, 68 (3) (1984), 665-70.
10.	H. Conrad, "Effects of electric current on solid state phase transformations in metals," *Mater. Sci. Eng. A*, 287 (2000), 227-237.

11.   P. Asoka-Kumar et al., "Detection of current-induced vacancies in thin aluminum-copper lines using positrons," *Appl. Phys. Lett.*, 68 (1996). 406-408.
12.   N. Bertolino et al., "High-flux current effects in interfacial reaction in Au-Al multilayers," *Philos. Mag. B*, 82 (2002), 969-985.
13.   A. Navrotsky et al., "Thermochemical insights into rapid solid-state reaction synthesis of β-sialon," *J. Phys. Chem. B*, 101 (46) (1997), 9433-9435.
14.   M.J. Tracy, and J.R. Groza. "Nanophase structure in Nb rich-Nb_3Al alloy by mechanical alloying," *Nanostructured Materials*, 1 (5) (1992), 369-378.
15.   V.A. Fedorov, Ya.A. Kesler, and E.G. Zhukov, "Magnetic Semiconducting Chalcogenide Spinels: Preparation and Physical Chemistry," *Inorg. Mater*. 39 (Suppl. 2) (2003), S68-S88.
16.   V.V. Tsurkan, private communication with author, Institute of Applied Physics – Academy of Science of Moldova, 2006.
17.   V.M. Indosova, E.G. Zhukov, and V.T. Kalinnikov, "The $FeS-Cr_2S_3$ System," *Inorg. Mater.*, 27 (1982) 533-536.
18.   D. Pearlman, E. Carnall, Jr.. and T.W. Martin, "The Preparation of Hot-Pressed Chalcogenide Spinels," *J. Solid State Chem.*, 7 (1973), 138-148.
19.   L.A. Stancin et al., "Electric-field effects on sintering and reaction to form aluminum titanate from binary alumina-titania sol-gel powders," *J. Am. Ceram. Soc.*, 84 (5) (2001). 983-985.

Materials Processing Under the Influence of External Fields

Session III

Materials Processing under the Influence of External Fields
Edited by Qingyou Han, Gerard Ludtka, Qijie Zhai
TMS (The Minerals, Metals & Materials Society), 2007

THE USE OF POWER ULTRASOUND FOR MATERIAL PROCESSING

Qingyou Han

Oak Ridge National Laboratory, Oak Ridge, TN 37831, USA

Keywords: Ultrasonic vibration, Degassing, Grain refinement, Welding, and Solidification

Abstract

The U.S. Department of Energy and Oak Ridge National laboratory have been sponsoring a number of projects investigating the use of high intensity ultrasonic vibrations for material processing. High intensity ultrasonic vibrations have been tested for degassing of molten aluminum, grain refinement of alloys of industrial importance, and modification of welding structure. The project results have indicated that power ultrasound can be used for fast degassing of molten metal, for significant grain refinement of alloys during their solidification processes, and for improving the microstructure in the weldment of steels. This article describes the project results and outlines the limits of using ultrasonic vibrations for material processing.

Introduction

The injection of high-intensity ultrasonic energy in liquid gives rise to nonlinear effects such as cavitation, acoustic streaming, and radiation pressure. Cavitation, or the formation of small cavities in the liquid, occurs as a result of the tensile stress produced by an acoustic wave in the rarefaction phase. These cavitation cavities continue to grow by inertia until they collapse under the action of compressing stresses during the compression half-period, producing high-intensity shock waves in the fluid. Acoustic streaming is a kind of turbulent flow that is developed neat various obstacles (interfaces) due to energy loss in the sound wave. These nonlinear effects can be utilized for the processing of molten metals and for the improvement of the solidification structure for ingots and castings.

In the past half century high intensity ultrasound has received much attention by scientists all over the world, especially in the former Soviet Union. Numerous applications have been tested including, for instance, ultrasonic cleaning, welding, machining, ultrasonic atomization, and metalworking. Ultrasonic treatment of aluminum alloys, in general, has been studied extensively [1,2]. It has been shown that the injection of high intensity ultrasonic vibration into aluminum alloy during its solidification can eliminate columnar dendritic structure, refine the equiaxed grains, and under some conditions, produce globular non-dendritic grains. Mechanisms for grain refinement under ultrasonic vibrations have been proposed but they are still arguable.

Recently, The U.S. Department of Energy and Oak Ridge National laboratory have been sponsoring a number of projects at the Oak Ridge National Laboratory and the University of Tennessee at Knoxville investigating the use of high intensity ultrasonic vibrations for material processing, including degassing of molten aluminum, grain refinement of alloys of industrial importance, and modification of microstructure of the weldment of steels. This article describes the project results and outlines the limits of using ultrasonic vibrations for material processing.

Ultrasonic Degassing

One of the applications of the ultrasonically induced cavitation phenomenon is degassing of molten metals. For some of the alloys of industrial importance, such as aluminum alloys, copper alloys, and steels, the solubility of dissolved gas in the liquid is much higher than that in the solid [3]. As a result, the dissolved gas precipitates from the liquid during solidification and forms pores among solid dendrites and grains. Porosity is one of the major defects in aluminum shape casting. The presence of porosity is detrimental to the mechanical properties, notably the fatigue resistance and the pressure tightness of the casting. There are a few degassing methods available. Rotary degassing is the most widely used method for degassing in aluminum alloys. During rotary degassing, a mixture of argon and chlorine is introduced into the molten metal through a specially designed graphite rotor so that bubbles can be broken and dispersed into the molten metal. The process is efficient but the use of chloride and moving part such as the graphite rotor is a drawback. High-intensity ultrasonic vibrations have a potential to be developed into a clean method for degassing.

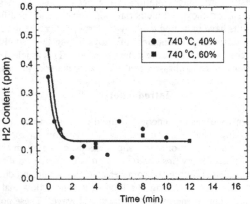

Figure 1: The hydrogen content as a function of ultrasonic processing time in melts prepared at two levels of humidity, 40% and 60% respectively.

An ultrasonic wave propagating through a liquid metal generates alternate regions of compression and rarefaction. When the intensity of the ultrasonic vibration is high enough, a large number of tiny cavitation bubbles is generated in the melt. The dissolved gas diffuses into the bubbles and is removed out of the melt as the bubbles escape from the melt surface. Figure 1 shows the hydrogen content as a function of times in the molten A356 aluminum alloy subject to high intensity ultrasonic vibrations [4]. Figure 2 illustrates the porosity level in the specimen obtained by reduced pressure testing (RPT). The experiment was carried out using a crucible containing 0.2 kg of aluminum melt at 740°C under two levels of humidity. Without ultrasonic vibration, the hydrogen content of the specimen cast under a humidity of 60% was much higher than that cast under a humidity of 40%. The initial hydrogen concentration was 0.45 and 0.36 ppm at humidity levels of 40 and 60% respectively. The RPT specimen was porous as it is shown in Figure 2(a). With ultrasonic vibrations, the hydrogen contents decreased sharply with increasing ultrasonic processing times in just the first minute. No much porosity can be seen on the polished section of the RPT specimen shown in Figure 2(b). With further increase in vibration times, the hydrogen content was further decreased until a steady-state hydrogen concentration was reached when the rate of hydrogen removal equaled to the hydrogen absorption by the molten metal. The results shown in Figures 1 and 2 suggest that ultrasonic degassing in a small aluminum melt is extremely fast. No matter what the initial hydrogen concentrations are, degassing can be achieved within one minute.

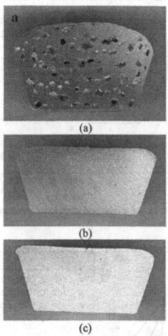

(a)

(b)

(c)

Figure 2: Sections of the reduced pressure test samples after different ultrasonic processing time under humidity 60% at melt temperature 740°C(The quality of the melt is 0.2 kg) (a) Without ultrasonic degassing; (b) Processing time = 1 min; (c) Processing time = 4 min

Ultrasonic degassing is sensitive to the metal temperature [4]. Figure 3 shows the efficiency of ultrasonic degassing in an A356 alloy melt under various melt temperatures. The results were obtained in a crucible containing 0.2 kg aluminum alloy. As illustrated in Figure 3, it took about one minute of ultrasonic vibration for the melt to reach a steady-state hydrogen content plateau when the melt was ultrasonically processed at a temperature of 700°C or 740°. The processing time required to degas the melt (to reach the steady-state hydrogen content plateau) increased with decreasing melt temperature in the temperature of 620 to 700 °C. It took almost 10 minutes to degas the melt held at 620°C, much longer than in the melt held at temperatures higher than 700°C.

One of the limitations for ultrasonic degassing is the volume of the melt. Figure 4 shows the degassing rates of ultrasonic vibration for various volumes of the melt. The experiments were carried out in melt at 700°C under a humidity of 60%. The weights of the melts were 0.2 kg, 0.6 kg and 2.0 kg, respectively. As illustrated in Figure 4, the ultrasonic processing time required for reaching the steady-state plateau hydrogen concentration increases with increasing volume (weight) of the melt. Degassing times were 1 min. for a 0.2 kg melt, 4 min. for a 0.6 kg melt, and 7 min. for a 2.0 kg melt. Degassing speed in a larger volume melt is much slower than that in a smaller melt. Fast degassing of a large-volume melt can be obtained by use of multiple ultrasonic radiators to inject ultrasonic vibrations into the melt. An innovative way of fast degassing a large volume is combining ultrasonic degassing with argon degassing. ORNL has an on-going project investigating degassing of a large volume aluminum alloy using ultrasonic vibrations.

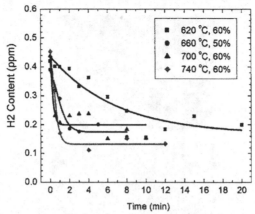

Figure 3: The hydrogen content as a function of ultrasonic processing time in melt degassed at different processing temperatures.

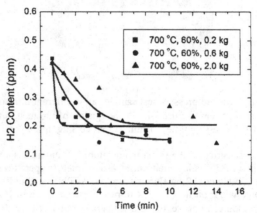

Figure 4: The hydrogen content as a function of ultrasonic processing time in melt of different sizes.

Grain Refining

Grain refining using Al-Ti-B master alloys have been widely used in the aluminum metalcasting industry. It is efficient but under some conditions leads to defect formation. Also it is undesirable to use grain refiners for pure aluminum. Grain refining of alloys using high intensity ultrasonic vibration has been studied extensively [1,2]. It has been demonstrated that the introduction of high intensity ultrasonic vibration into the melt can refine the equiaxed grains, and under some conditions, produce globular non-dendritic grains [1]. To conform these results and to better evaluate the effect of ultrasonic vibration on the microstructure formation during solidification of alloys, projects were funded by the US Department of Energy and Oak Ridge National Laboratory. These projects evaluated the methods and conditions for achieving the maximum effect of graining refining using high-intensity ultrasonic vibrations.

100

Figure 5 shows grain refinement in aluminum A356 alloy [5]. The alloys were cast into a copper mold holding a specimen about 5 cm diameter hemisphere. Without ultrasonic vibrations, primary aluminum grains were dendritic and the grain sizes are large. With ultrasonic vibrations, the primary aluminum grain sizes were reduced. When the ultrasonic vibration was injected into the ingot under optimum conditions, spherical grains in the size range of 30μm could be obtained. 30μm might be the smallest grain size that has never been produced using other grain refining techniques for such a large ingot. These spherical grains are essentially suitable for the semi-solid processes.

The optical micrographs of the silicon phase in A356 alloy treated with and without high-intensity ultrasonic vibration are also shown in Figure 5. Samples without subject to ultrasonic vibration exhibited coarse acicular eutectic silicon dispersed among the fully developed primary aluminum dendrites. The eutectic silicon was about 100 μm in length. In contrast, the ultrasonically treated A356 alloy displayed a microstructure of continuous very fine eutectic silicon interspersed with fine globular primary aluminum grains. The shape and size of individual eutectic silicon grains were undistinguishable at such a magnification.

(a)   (b)

Figure 5: The grain size of aluminum A356 alloy cooled in a copper mold. (a) without ultrasonic vibration, and (b) with ultrasonic vibration for 20 s (20 kHz and 750 W).

Figure 6 shows the comparison of the 3-dimensional morphology of eutectic silicon observed by SEM on the deep-etched samples [6]. The eutectic silicon with ultrasonic treatment exhibited a rosette-like form; whereas it had a typical coarse plate-like form without ultrasonic treatment. The result again suggested that the modified eutectic silicon is not simply a finer form of the unmodified A356 alloy. The mechanism by which the finer eutectic silicon phase forms under power ultrasound is not clear at the moment. However, the rosette-like structure of the silicon phase shown in Figure 6 (b) is an indication that the eutectic cells were nucleated in the remaining liquid among dendrites during eutectic solidification. The center of the rosette could be the center of a eutectic cell. The untreated specimen exhibited plate-like structure which was most likely due to the nucleation and the subsequent growth of the eutectic cells on the existing aluminum dendrites (shown in Figure 6 (a)), rather than the independent nucleation and growth of the eutectic cells in the liquid (Figure 6(b)).

The mechanisms of grain refining using high-intensity ultrasonic vibration were also investigated in these projects. Ultrasonic vibrations were injected into the A356 alloy at various temperatures both higher and lower than the liquidus temperature of the alloy [5]. The results indicated that when the ultrasonic vibration was introduced into the alloy at mushy zone temperatures, the primary aluminum grains could be refined slightly but no spherical grains could be obtained during normal solidification times. The best grain refining effect was obtained when the ultrasonic vibration was introduced in the molten alloy about 10 °C higher than the liquidus temperature. This is a clear indication that the grain refining is mainly related to the nucleation stage of the solidification rather than the fragmentation of dendrites. I believe that a large number of nuclei are formed on the cavitation bubbles during their growth stage (during the rarefaction phase of the ultrasonic vibration) and are dispersed into the molten metal when these cavitation bubbles are collapsed during the compressive phase of the vibrations.

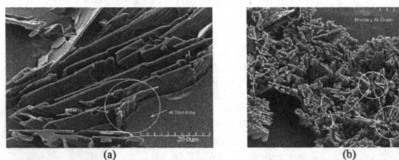

(a)   (b)

Figure 6: Three-dimensional eutectic morphology observed using SEM at 2000X magnification. (a) without ultrasonic treatment and (b) with ultrasonic treatment.

Welding

When considering the operational performance of weldments in the engineering projects, the most important issues to be considered are weld metal mechanical properties, integrity of the welded joint, and weldability [7]. These issues are closely related to the microstructure of the weld metal. A significant amount of research has been carried out to alter the process variables and to use external devices to obtain microstructure control of the weldments. It has been reported that grain refined microstructure not only reduces cracking behavior of alloys including solidification cracking, cold cracking and reheat cracking [8], but also improves the mechanical properties of the weld metal, such as toughness, ductility, strength, and fatigue life. Weld pool stirring, arc oscillation, arc pulsation, and magnetic arc oscillator have been applied to fusion welding to refine the microstructures. We have tried using high-intensity ultrasonic vibration to modify the microstructure of the weldments.

Figures 7 and 8 show the typical microstructures of the weld deposits extracted from the transverse sections of specimens produced with and without using ultrasonic vibrations [9]. These photomicrographs consist of several figures taken continuously along the fusion line (FL) from the top to the bottom of the cross sections. Figure 7 shows the typical weld metal microstructure. columnar grains were produced without using ultrasonic vibration. During solidification process, grain growth initiated from the liquid phase on the existing crystalline substrate at the fusion boundary and proceeded toward the weld centerline, therefore extending it without altering the crystallographic orientation. Figure 8 shows the microstructure produced by using ultrasonic vibration. The columnar dendritic microstructure was substantially reduced in comparison with those in Figure 7. Equiaxed grains formed due to high intensity ultrasonic vibrations.

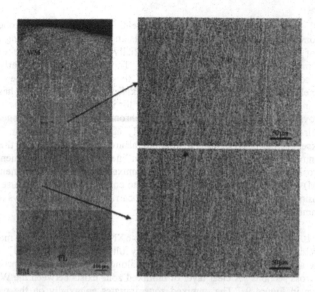

Figure 7: Microstructure of commercial 316L weld metal without using ultrasonic vibration.

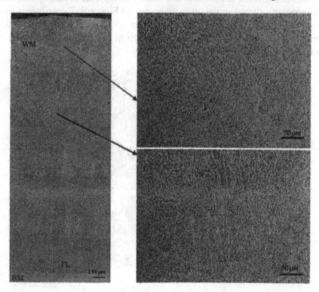

Figure 8: Microstructure of commercial 316L weld metal subject to ultrasonic vibration.

Another application of ultrasonic vibration is for the elimination of unmixed zone in the weldment of super-austenitic stainless steels.

Unmixed zone is a boundary layer adjacent to the fusion boundary and consists of base metal which is melted and solidified during welding without experiencing mechanical mixing with the filler metal [10]. As a fundamental phenomenon in welding of steels, unmixed zone is thought to occur for all combinations of filler and base materials and for all fusion welding methods. The solidification of the unmixed zone initiates epitaxially on the partially-melted region and proceeds in a manner identical to the fusion zone. This region may be clearly observed in heterogeneous couples where the melting range of the consumable electrode approached or exceeded the melting range of the base metal. The existence of unmixed zone has been caused more concerns in recent years in the welding industry for its effect on weld-associated corrosion. Very often, alloys with high levels of molybdenum, chromium, nickel, and nitrogen are required to increase their corrosion resistance in a chemically aggressive environment. However, no matter how excellent the base metal and filler metal are individually, when they are welded together, the unmixed zone will form along the fusion line in a solidified weldment. It has been found that corrosion can preferentially occur in the unmixed zone in an environment in which the base metal (BM) and weld metal (WM) are resistant to corrosion [11,12]. Because of its inferior localized corrosion resistance, great efforts have been carried out to find the ways to decrease the width of the unmixed zone.

Figures 9 and 10 show the microstructure of an AL-6XN SMAW BOP weld made with C-22 filler metal welded with and without high intensity ultrasonic vibration [13]. These figures consist of several micrographs taken continuously along the fusion line from the top to the bottom of the cross sections. A clear layer of unmixed zone exists between the WM and BM of the weld shown in Figure 9. The unmixed zone initiates epitaxially on the partially-melted region and proceeds in a manner identical to the fusion zone. This is clearly indicated in the high magnification micrograph shown in the up right corner of Figure 9. Experiments have shown that fluid flow or convection in the weld is not sufficient to eliminate the unmixed zone since a laminar boundary layer always exists at the freezing front. In this laminar layer, mixing of the metals is controlled by diffusion in the liquid.

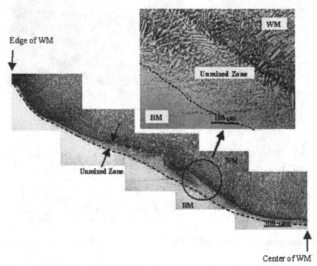

Figure 9: The distribution of the unmixed zone along welding interface on transverse cross-section of AL-6XN SMAW BOP weld (C-22 filler meal) without using ultrasonic vibration. An unmixed zone exists between the base metal (BM) and the weld metal (WM).

Figure 10 shows the microstructure of the weld subjected to ultrasonic vibrations throughout the welding process. No unmixed zone formation was observed along the entire fusion line. This can be attributed to a thorough mixing between the melted BM and WM under the influence of ultrasonic vibration. The injection of high-intensity ultrasonic fields in liquid gives rise to nonlinear effects such as cavitation and acoustic streaming. Both of these nonlinear effects are strongest near the solid-liquid interface during the melting and solidification of the weld. In fact, the ultrasonically induced turbulent flow is initiated at the solid-liquid interface adjacent to the location where the unmixed zone usually occurs. This turbulent ultrasonically induced flow provides strong mixing of the molten base metal with the molten filler metal, resulting in the eliminating of the unmixed zone during the welding of the super-austenitic stainless steel.

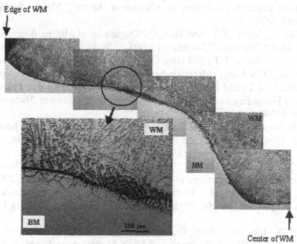

Figure 10: Photomicrograph of welding interface on transverse cross-section of AL-6XN SMAW BOP weld with ultrasonic vibration. No unmixed zone is formed between the base metal (BM) and the weld metal (WM).

Conclusions

Research that has been carried out at Oak Ridge National Laboratory and the University of Tennessee at Knoxville indicates that high intensity ultrasonic vibration can be used for a rapid degassing of molten aluminum alloys, and for modification of the solidification structure such as the size and morphology of the primary phase and the secondary phase during the solidification process of the alloy. Small and spherical grains in the size range of 30 µm can be obtained in aluminum A356 alloy about 2 in. hemisphere. The columnar structure in the stainless steel weldment can also be altered to equiaxed structure. Furthermore, the unmixed zone in the super-austenitic stainless steel weld can be eliminated. However the beneficial effect of ultrasonic vibration on alloy process is mainly achieved in small specimens. Further effort needs to be made to extend these research results into industrial applications.

Acknowledgment

This research was supported by the United States Department of Energy, Assistant Secretary for Energy Efficiency and Renewable Energy, Industrial Technologies Program, Industrial Materials for the Future (IMF), Aluminum Industry of the Future, under contract No. DE-FC36-

02ID14399 with UT-Battelle, LLC. The author would like to thank Drs. T.T. Meek, H. Xu, X. Jian, Y. Cui, and C. Xu, and Mr. E. Hatfield for their help in the experiments on degassing, grain refining, and welding of alloys.

References

1. Abramov, O.G., "High-Intensity Ultrasonics, Gordon and Breach Science Publishers, Amsterdam, The Netherlands, 1998.
2. Rozenberg, L.D., (ed.), Physical Principles of Ultrasonic Technology – vol. 2, Plenum Press, New York, 1973.
3. Han, Q., and Viswanathan, S., "Hydrogen Evolution during Directional Solidification and Its Effect on Porosity Formation in Aluminum Alloys," Metall. Mater. Trans., 33A (2002), 2067-2072.
4. Xu, H., Jian, X., Meek, T.T., and Han, Q.,"Degassing of Molten Aluminum A356 Alloy Using Ultrasonic Vibration," Materials Letters, v 58/29, 2004, pp. 3668-3672.
5. Jian, X., Xu, H., Meek, T.T., and Han, Q., "Effect of Power Ultrasound on Solidification of Aluminum A356 Alloy," Materials Letters, vol. 59, issues 2-3, 2005, pp. 190-193.
6. Jian, X., Meek, T.T., and Han, Q., "Refinement of Eutectic Silicon Phase of Aluminum A356 Alloy Using High-Intensity Ultrasonic Vibration," *Scripta Materialia.*, 2006, vol. 54, n5, pp. 893-896.
7. Folkhard, E., Welding Metallurgy of Stainless Steels, Springer-Verlag Wien New York, 1984.
8. Easterling, K., Introduction to the Physical Metallurgy of Welding, Butterworths, 1983, pp. 73.
9. Cui, Y., Xu, C.L., and Han, Q., "Microstructure Improvement in Weld Metal Using Ultrasonic Vibrations" Submitted to Advanced Engineering Materials, 2006.
10. Karlsson, K., and Arcini, H., "Formation of Unmixed Zones in Stainless Steel Welds," presented at Int. Conf. on Innovation of Stainless Steel, Florence, Italy 11-14 Oct. 3.273 (1993).
11. Lundin, C.D., Liu, W., Zhou, G., and Qiao, C.Y., "Unmixed Zone in Arc Welds: Significance on Corrosion Resistance of High Molybdenum Stainless Steels," Welding Research Council bulletin, 428, 1998, pp. 1-98.
12. Matthews, S.J., and Savage, W.F., "Heat-Affected Zone Infiltration by Dissimilar Liquid Weld Metal," Welding Journal. 71, 1971, pp.174s-182s.
13. Cui, Y., Xu, C.L., and Han, Q., "A Method of Eliminating the Unmixed Zone in AL-6XN Alloy Welds," Scripta Materialia., 2006, vol.55, No.11, pp. 975-978.

Materials Processing under the Influence of External Fields
Edited by Qingyou Han, Gerard Ludtka, Qijie Zhai
TMS (The Minerals, Metals & Materials Society), 2007

INFLUENCE OF POWER ULTRASONIC ON SOLIDIFICATION STRUCTURE AND SEGREGATION OF 1CR18NI9TI STAINLESS STEEL

Qing-Mei LIU[1], Yu-Lai GAO[1], Fei-Peng QI[1], Qi-Jie ZHAI[1], Qing-You HAN[2]

[1]Center for Advanced Solidification Technology (CAST), Shanghai University, Shanghai, 200072, P.R. China
[2]Oak Ridge National Laboratory, Oak Ridge, TN 37831-6083, U.S.A.

Key words: Ultrasonic treatment, Grain refinement, Dendritic segregation

Abstract

The present investigation evaluates the effect of high-intensity ultrasonic treatment (UT) on the solidification behaviors of 1Cr18Ni9Ti austenitic stainless steel. The melt is treated with 600W UT by self-developed ultrasonic-introduced system during solidification. In comparative with without UT, after 600W UT, not only the macrostructure is refined remarkably, but also the morphology of the dendrite structure is improved from coarse long dendrite into fine equiaxed one. EDS result shows that the microsegregation is alleviated with UT. Based on experimental results, the mechanism for UT on the melt is discussed.

Introduction

Since the middle of last century, some researches have been carried out to apply high-energy ultrasonic treatment on the melt during solidification and made pronounced effects on refining structure, degassing and eliminating microsegregation [1-4]. However, most of former researches were focused on alloys with low melting temperatures [5-10]. Although steels are widely used in the industry, there are few papers to make in-depth analysis of steels by treated with ultrasonic due to the difficult operation of this process on high melting temperature alloys.

Hence, in this paper, we propose small-scale model experiments using self-developed ultrasonic-introduced system for detailed investigations of the influence of UT on the steel during solidification. Macrostructure, microstructure and microsegregation without and with 600W UT are comparably investigated. Based on experimental results, the influence mechanism of UT during solidification is discussed.

Experimental details

Schematic illustration of experimental setup is shown in Figure1. The ultrasonic source devices are composed of an ultrasonic power, an ultrasonic generator with the frequency of 20kHz and an electromagnetic amplitude transformer horn. Along the horizontal direction, the electromagnetic amplitude transformer horn is inserted into the mould from the profile. The amplitude and vibration intensity of the horn are directionally proportion to the outputting power of ultrasonic generator. The composition of the austenite stainless steel used in this study is presented in Table1. Material was melted by a medium frequency induction furnace. After deoxidization with Al at 1650 °C, the melt was poured into a sand mould with heat insulating material on the mould wall. Sand mould had a dimension of 60mm×60mm×120mm. Before melt poured into the mould, the horn and mould were preheated to 500°C. Two groups of cast samples were prepared under the same casting conditions; one is treated with 600W UT during solidification and another without UT. For convenience, without UT was defined as 0W UT, too. After solidification, two casts cooled to room temperature. Two samples, used in macrostructure and microstructure examinations, were polished and etched with $15gFeCl_3+50mlHCl+50mlalcohol$. Scanning electron microscope (SEM) observations of microstructure and energy dispersive spectroscopy (EDS) investigation were performed using JSM-6700F.

Table 1. Composition of the austenite stainless steel (wt. %)

C	Si	Mn	Cr	Ni	Ti	S	P	Balance
0.06	0.56	0.99	17.45	8.96	0.41	<0.04	<0.03	Fe

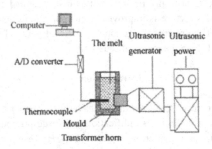

Figure 1.Schematic illustration of experimental

Results and discussion

Effect of UT on structure of steel

The influence of without and with UT on grain refining performance of 1Cr18Ni9Ti is shown in Figure2. Without UT, the structural morphology of this steel is typically coarse columnar grain, as shown in Figure2 (a). Macrostructure with 600W UT shows the grain morphology becomes fine and the average grain space is decreased significantly. According to the presented results, it can be concluded that the grain refining performance of 1Cr18Ni9Ti treated by power ultrasonic is improved.

Figure3 is microstructure of 1Cr18Ni9Ti steel in both conditions. The microstructure consists of austenite matrix with a little ferrite phase at dendrite arms. Figure3 (a) and (b) display a large

108

difference in terms of the dendrite size. Without the application of UT, it exhibits long column dendrite with developed primary dendritic arm. However, Figure3 (b) reveals that long dendrite can successfully be broken into pieces by power ultrasonic and transform into finer equiaxed one.

When power ultrasonic is introduced into a liquid medium, because the external applied ultrasonic is sinusoidal acoustic wave, a strong nonlinear effect will be generated. The nonlinear effect causes the cycling alternating positive-negative pressure in the local zone of the melt. During the negative period of the cycle, if the negative pressure of ultrasonic is larger than the summation of environment pressure and liquid strength, some small cavitation bubbles appear at weak regions of the melt. Then cavitation bubbles expand their sizes many times instantaneously and if the intensity of positive pressure is high enough, cavitation bubbles can collapse rapidly in the following positive period of the cycle. As a result, cavitation effect resulted in the fluctuation of temperature and pressure occurs in the liquid. During the expansion stage of cavitation bubbles, the metal temperature on the bubbles surfaces drops because bubbles absorb energy from the near liquid metal and this gives rise to the formation of lots of nuclei [11]. Additionally, these extrinsic nuclei can uniformly distribute in the whole melt with the intense stirring effect of acoustic streaming. Thus, the grain refining performances of the steel is evidently improved after UT.

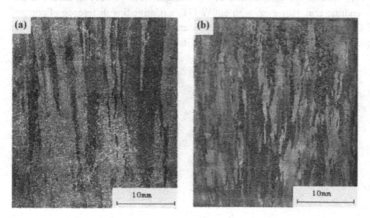

Figure 2. Macrostructure of 1Cr18Ni9Ti stainless steel without (a) and with (b) 600W UT.

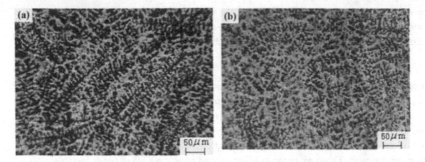

Figure 3. Microstructure of 1Cr18Ni9Ti stainless steel without (a) and with (b) 600W

Effect of UT on microsegregation

Figure4 shows SEM images together with the EDS points of the samples in both conditions and the corresponding composition distributions of Cr at different points are given in Table 2. The content distribution of Cr in austenite matrix and dendritic arms shows that without the application of UT, the average composition of Cr changes from 17.86% in austenite grain to 55.32% in the dendritic arm. However, with 600W UT, a relative tiny difference of composition distribution of Cr is observed between matrix and dendritic arm. ESD results indicate that the amplitude of microsegregation in the sample with UT is much less than that without UT.

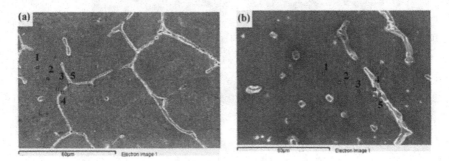

Figure 4. SEM images together with the EDS points of the samples without (a) and with (b) 600W

Table 2. The corresponding composition distributions of Cr at different points (wt.%)

Power (W)	1	2	3	4	5
0	17.86	18.22	16.98	55.22	55.60
600	18.31	17.38	18.05	28.48	28.57

The reason that UT can alleviate microsegregation is due to the fact that when cavitation bubbles collapse, local high temperature is generated. The high temperature can inflect the short-distance diffusion remarkably in the front of the solid-liquid surface. According to the diffusion law, the

110

diffusion coefficient strongly depends on local temperature and the relation between diffusion coefficient D and temperature T is [12]:

$$D = D_0 \cdot \exp\left(\frac{-E}{RT}\right)$$ (1)

where D_0 is diffusion constant, E diffusion activation energy, R gas constant.

During UT, the temperature of local zone generated by cavitation bubbles is:

$$T = T_0 \frac{P(\gamma - 1)}{Q\gamma}$$ (2)

where T_0 is the vapor temperature in cavitation bubble, i.e.373K. P is the environment pressure, i.e. 1atm. Q is the pressure in bubble, γ ratio of specific heat of vapor, i.e. 1.3.

Combined (1) and (2), we can get:

$$D = D_0 \cdot \exp\left(\frac{-EQ\gamma}{RT_0 P(\gamma - 1)}\right)$$ (3)

Let Q be 0.01atm, we can calculate that the instantaneous temperature in local zone is 8607K. Due to the energy fluctuation in local zone, Cr atoms overcome potential barrier and transport, causing the improvement of short-distance diffusion. Thus, the increase of diffusion velocity in the phase interface benefits to distribute Cr atoms uniformly and alleviate microsegregation.

Conclusions

1. UT has a significant effect on refining the macrostructure from coarse column grain to fine one. In addition, the morphology of microstructure is modified from developed columnar dendrite without UT to finer equiaxed dendrite with 600W UT.
2. With the application of power ultrasonic into the melt during solidification, the microsegregation between austenite matrix and dendritic arm is alleviated remarkably.

Acknowledgement

The authors would like to thank the National Natural Science Foundation of China (Grant No. 50374046, 50574056) for financial support.

References

[1]. Abramov, *Ultrasound in Liquid and Solid Metals* (New York, NY: CRC Press, 1994).

[2]. G.I. Eskin, *Ultrasonic Treatment of Light Alloy Melts* (Amsterdam: Gordon&Breach Press, 1998).

[3]. H. Xu, X. Jian and Q. Han, "Ultrasonic Degassing of Molten Aluminum under Reduced Pressure," *TMS Light Metals*, (2005), 915-919.

[4]. Y.B. Choi et al., "Effect of Ultrasonic Vibration on Infiltration of Nickel Porous Preform with Molten Aluminum Alloys," *Materials Transactions*, 46 (2005), 2156-2158.

[5]. V.I. Slavov et al., "Effect of Shock Loading with Ultrasound on the Structure and Properties of the Metal of Welds and Products," *Metallurg*, 6 (2005), 63-65.

[6]. J. Li, T. Momono, "Effect of Ultrasonic Output Power on Refining the Crystal Structures of Ingot Sand Its Experimental Simulation," *Journal of Materials Science and Technology*, 21 (2005), 47-52.

[7]. G.N. Kozhemyakin and L.G. Kolodyazhnaya, "Influence of Ultrasonic Vibrations on the Growth of InSb Crystals," *Journal of Crystal Growth*, 149 (1995), 267-271.

[8]. V. Abramov et al., "Solidification of Aluminum Alloys under Ultrasonic Irradiation Using Water-cooled Resonator," *Materials Letters*, 37 (1998), 27-34.

[9]. H. Xu et al., "Degassing of Molten Aluminum A356 Alloy Using Ultrasonic Vibration," *Materials Letters*, 58 (2004), 3669-3673.

[10]. X. Jian, T.T. Meek and Q. Han, "Refinement of Eutectic Silicon Phase of Aluminum using High-intensity Ultrasonic Vibration," *Scripta Materialia*, 54 (2006), 893–896.

[11]. X. Jian et al., "Effect of Power Utrasound on Solidification of Aluminum A356 Alloy," *Materials Letters*, 59 (2005), 190– 193.

[12]. Y.Y. Li et al., "Heat Treatment of 2024/3003 Gradient Composite and Diffusion Behavior of the Alloying Elements," *Materials Science and Engineering A*, 391 (2005), 124–130.

A NUMERICAL INVESTIGATION OF THE SOLIDIFICATION OF AL-12.6%SI ALLOY DURING MOLD-VIBRATION

Numan Abu-Dheir[1], Marwan Khraisheh[2], Kozo Saito[3]

[1]Mechanical Engineering Dept, King Fahd University of Petroleum & Minerals
Box 506, Dhahran, 31261, SA
[2,3]Center for Manufacturing, and Mechanical Engineering Department, University of Kentucky
Lexington, KY-40506-0108, USA
Tel: (859)-257-6262 ext 219 Fax: (859)-257-3304
[1]abudheir@kfupm.edu.sa, [2]khraisheh@engr.uky.edu, [3]saito@engr.uky.edu,

Keywords: Casting, Mold vibration, Numerical simulation, Mold-casting interface heat transfer coefficient, Solidification time.

Abstract

A simple 1-D numerical model to predict the effect of mold-vibration on the solidification of metals is presented. Mold-vibration during solidification is known to have profound effect on the microstructure and mechanical properties of castings. However, its working mechanism is not well understood yet. Inverse heat conduction method is used to estimate mold/casting heat transfer coefficient, and phase-change during the solidification of Al-12.6%Si. The results for different conditions of vibration are compared. Numerical simulation results show that mold-vibration has changed the heat transfer coefficient in a way dependant on the frequency and amplitude used. Mold-vibration also assisted the phase-change process, by reducing nucleation time and in some cases reducing the solid growth time.

Introduction

In a recent study we have seen that mold vibration has profound influence on the casting process [1-2]. It was proven that mold-vibration results in the refining of the solidified microstructure of Al-12.6%Si alloy. Mold vibration was also found to result in a change in the temperature distribution inside the mold. Depending on these results we concluded that mold vibration plays a significant role in altering the solidification kinetics, mainly seen by the change on the solidification time. As we want to provide more detailed explanation this effect, the current work will use solidification macroscopic modeling to generate temperature profiles, phase change, and solidification time for the casting made under the presence of mold-vibration.

Mechanical vibration as a technique for grain refinement was first reported early in the last century [3]. Several following studies have since investigated the effect of vibration on the microstructure of castings [1-2, 4-6]. The beneficial effects of vibration were observed with several types of metals, e.g. zinc, brass, aluminum, etc. The effects include promotion of nucleation and thus reducing as-cast grain size, reducing shrinkage porosities due to improved liquid feeding, and producing a more homogenous metal structure. These improved features led to enhanced mechanical properties and lower susceptibility to cracking. However, the lack of quantitative conclusive correlation between vibration parameters and the resulting refinement has prevented this technique from being more widely used. In order to develop a quantitative model to predict the effect of vibration on the produced casting, the effect of the vibration parameters should be known. Two mechanisms explaining this effect have been reported. The first mechanism suggests that vibration would enhance the creation of cavities inside the solidifying metal and aiding it to grow so it would eventually collapse resulting in an enormous exhausting of pressure wave that shatters the formed grains to form a new nucleation sites. The second theory states that the shear force generated from the vibrating fluid will break the weak dendrite arms and the broken arms will increase the nucleation sites. Both mechanisms, however, suggest that applying vibration increases the number of nucleation sites to refine the microstructure. In an earlier study [1] the temperature distribution inside the mold was measured and it was found that mold temperature was greatly influenced by the mold vibration. Therefore, more understanding on the effect of vibration on the solidification process is needed.

Different methods have been used to analyze the solidification process. Few of these were analytical [7, 8], many experimental [9-12], and an increasing number of numerical methods. Solidification of metals is a very complex process and analytical methods alone do not provide enough insight on the characteristics of the output casting. At the same time experimental techniques help in analyzing the processes under investigation, but it can not provide many answers on how to optimize the process and it is difficult to apply in case where rapid cooling for the casting is involved. Thus, numerical methods have been used extensively to gain more understanding of the metal solidification. More methods to describe the microstructure were provided by Stefanescu [13], where he reviewed and gave comparison of the different microscopic models in the literature

The heat transfer coefficient between the mold and the casting, $h_{mold/casting}$, is an important factor for the solidification process. It determines the manner by which heat will transfer from the casting to the mold and thus directly affecting the casting solidification. While uniform temperatures gradients can exist in both metal and mold, the junction between the two surfaces creates a temperature drop, which is dependent upon the thermophysical properties of the contacting materials, the casting and mold geometry, the roughness of mold contacting surface, the presence of gaseous and non-gaseous interstitial media, the melt superheat, contact pressure and initial temperature of the mold [14]. The uncertainty associated with evaluating the heat transfer coefficient due to the difficulty in measuring the surface temperatures and the nonlinearity of its mathematical formulation made finding this term especially hard and a topic for academic research. Usually, researchers use an estimated value of the heat transfer coefficient [15], or use the inverse heat conduction problem [16] which is most popular for evaluating the value of $h_{mold/casting}$. Prates et al [17] concluded that the value of the heat transfer coefficient controls the number of nuclei formed in contact with the mold surface.

In the current work, we will employ a 1-D macroscopic numerical technique to simulate the effect of mold-vibration on the solidification characteristics of a eutectic Al-Si alloy, namely solidification characteristics of the casting/mold heat transfer coefficient, nucleation, and solidification time. This will be aided by developing time function for the heat transfer and monitoring the development of solid fraction inside the casting. As we want to provide more detailed explanation on how this is happening, the current work will use solidification macroscopic modeling to generate temperature profiles, phase change, and solidification time for the casting made under the presence of mold-vibration. The output results of the numerical simulation should be directly related to the vibration conditions at which the numerical simulations are conducted. Therefore, this study aims at shedding light on the mechanism by which mold-vibration influencing the solidification process.

Experimental Set Up

To produce a wide range of vibration conditions, a computer-controlled 500-lbf electromagnetic shaker is used to vibrate a steel mold. A prepared mix of Al-12.6%Si alloy is melted at 850°C and poured inside the vibrating mold, temporal temperature of the casting and the mold is recorded. More details of the experiment can be found elsewhere [1]. For the purpose of the inverse heat calculation, two thermocouples have been used inside the mold and one inside the casting as in Figure 1.

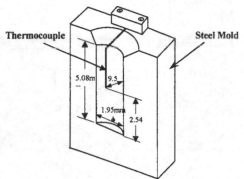

Figure 1. Cross-sectional view showing the
location of the thermocouple inside mold cavity

Model Development

Because of the complexity of the solidification process and the many parameters affecting the casting, it is impossible to develop an analytical model that can accurately describe the effects of vibration on solidification. On the other hand, it is not practical to rely only on experimental data to analyze and quantify the vibration parameters on casting. A compromise would be to use simple numerical technique that couple experimental data and an analytical model to develop a tool to learn and predict the solidification behavior due to vibration. Therefore, we choose to use inverse heat conduction analysis to develop such a model.

In macroscopic scale, the solidification is primarily controlled by heat diffusion for both the liquid and solid regions [18]. Neglecting the convective heat transfer, the radial 1-D governing equation [19]

$$\frac{k}{\rho c_p}\left[\frac{1}{r}\frac{\partial}{\partial r}\left(r\frac{\partial T}{\partial r}\right)\right]+\frac{\dot{q}}{\rho c_p}=\frac{\partial T}{\partial t} \tag{1}$$

Where k, c_p, r, T, t and ρ are in W/(m K), W/(m K), m, sec and °C, kg/m³ respectively. All of the values for the properties of the mold and casting can be found elsewhere [20, 21]. The 1-D heat transfer is justified by the strong effect of the high temperature gradient in the radial direction compared to the other directions. The high temperature difference between the superheated casting and the cold mold will ensure that the bulk heat flow will be unidirectional, as demonstrated in Figure 2. An adjusted value for the heat conduction coefficient to account for the heat convection has been used by others [22], however in the current case the heat convection of the small-sized casting is not of primary concern. All of the physical properties, for both the casting and mold, are listed in the appendix.

The term $\dot{q}$ in equation 1 is the rate of the released latent heat, which is a result of phase transformation and is expressed by [19]

115

$$\dot{q} = L \dot{f}(r,t) \qquad (2)$$

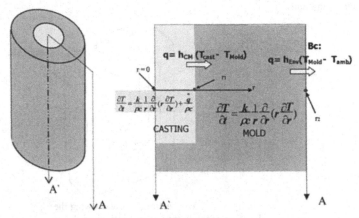

Figure 2. Heat flow representation in the cross section of a
cylindrical mold-casting system.

L in eq'n 2 is the latent heat, J/kg, of fusion resulting from phase-change and $f(r,t)$ is the solid fraction at a given time t and coordinate r. A simple relation between temperature and solid fraction is assumed and the rate of phase-change can be expressed as

$$\dot{f}(r,t) = \frac{df_s}{dT} \dot{T}(r,t) \qquad (3)$$

Geometry and Assumptions

The geometrical description for the casting and mold is an important aspect of the modeling process. Neglecting the presence of a cone at the opening of the mold, Figure 1, it was assumed that that casting has a completely of a cylindrical shape, as seen in Figure 2. The cylindrical geometry of the casting will make the heat to be omitted in the radial form inside the mold.

In addition, it essential to know that the forthcoming model will only consider conduction heat transfer inside the casting and neglect any convection since it is not of primary concern as discussed earlier.

The essence of the current work is based on the assumption that vibration influences the heat evolved from the phase transformation. This assumption is made due to fact that vibration will introduce energy to the solidifying system and thus will influence the mode of solidification. Thus, a vibration phase-change correlation term, α, is used to introduce the effect of vibration on phase change as follows

$$\dot{q} = L \dot{f}(r,t) \bullet \alpha \qquad (2^*)$$

116

The value of term α is 1 for the case of casting without vibration. Using the method of reverse heat calculation, this value will be modified as vibration is introduced in a manner proportional to the measured values of the casting temperature.

Substituting equations (3) and (2*) into equation (1) and rearranging we will get the following expression for temporal temperature variation

$$\dot{T}(r,t) = \frac{\text{div}[k \cdot grad(r,t)]}{\rho(c_p - L\alpha \frac{df_s}{dT})} \tag{4}$$

Three boundary conditions are used to asses in the modeling process; at the casting/mold interface, at mold/environment interface, and at the center of the casting.

Boundary Conditions

The first boundary condition is the average heat transfer at the interface between the mold and the casting. As a result the boundary condition at the mold/casting interface, Figure. 2, will be

at $r = r_1$ $\qquad\qquad$ $q_{cm} = h_{mold/casting} (T_C - T_M)$ $\qquad\qquad$ (5)

The unknown geometry of the contact surface and the difficulty associated with the measurement of the surface temperature of both the mold and the casting, led to the use of the inverse heat conduction method. This method is used to estimate the surface heat flux history given one or more measured temperature histories inside a heat-conducting body, or to estimate a surface condition from interior measurements [23].

The second boundary condition is at the mold-environment interface. Heat convection is assumed to be the dominant governing factor and the heat transfer coefficient is assumed to be constant at 10 W/m^2K with the environment temperature at 25°C. Thus

at $r = r_2$ $\qquad\qquad$ $q = 10 \ (W/m^2K) \ (T_M - 25°C)$ $\qquad\qquad$ (6)

Finally, symmetry is assumed to exist along the longitudinal axis of the casting. Thus

at $r = 0$ $\qquad\qquad$ $\frac{\partial T}{\partial r} = 0$ $\qquad\qquad$ (7)

The numerical simulation time step is calculated from the stability condition and found to be about 0.0001 second ($\Delta t = \frac{(\Delta r)^2}{2(k/\rho c)}$). The chart in Figure 3 details the steps followed in the numerical simulation.

Results and Discussion

Validation of results

Temperature Profiles. To validate the results of the numerical simulation, a comparison of the readings of thermocouple inside the casting and inside the mold and that of the calculated results of the correspondent node is made. As seen in Figure 4, the inverse heat calculations are most accurate among the computation nodes corresponding to thermocouple just inside the mold wall

117

and the thermocouple inside the cavity. Figure 4 also shows that the numerical solution was able to capture the physics of the solidification process. The closeness of numerical results of the casting temperatures and the corresponding experimental temperature during the solidification stage is mainly depending on the assumed solidification temperature range, which was chosen to be 568°C to 572°C. This temperature range is used for all castings regardless of the vibration conditions.

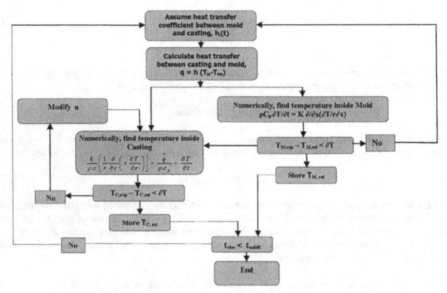

Figure 3. Schematic representation of the numerical simulation algorithm.

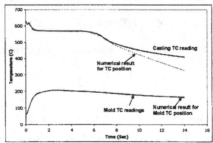

Figure 4. Plots of the thermocouple readings and the correspondent
numerical simulation results for the case of casting with vibration at 100 Hz and 49.

Numerical Solidification Time. Figure 5 shows the plots of the solidification time for castings with different vibration conditions. To get a closer view on the accuracy of the solidification numerical simulation, the solidification time measured from the experiment and that calculated from numerical simulation results are compared. The solidification time from the numerical results is taken as time from the point when temperature is equal to 572°C to the point when temperature reaches 568°C. It can be seen from Figure 5 that the solidification times calculated from the numerical model results has strong agreement with the measured solidification times. The underestimated values of the numerical solidification times is due to the fact that the

118

measurement of the experimental solidification time does not take into account the remelting process that occur at the beginning of the solidification process.

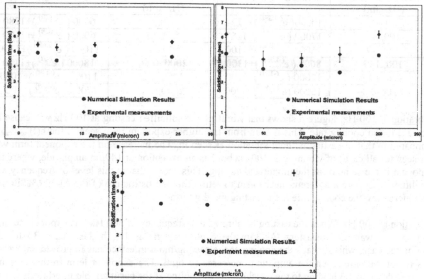

Figure 5. Solidification time vs. amplitude for casting solidified with vibration at 100Hz (top left), 500Hz (top right), 2kHz (bottom).

Inverse Heat Calculation Results

Mold-casting Interface heat transfer coefficient. At the start of the numerical simulation the heat transfer coefficient value is first assumed. Based on the accuracy of the developed temperature profiles computed from the inverse heat calculation of the mold the value of $h_{mold/casting}$ will be recalculated. Since $h_{mold/casting}$ is varying with time, the predicted value of $h_{mold/casting}$ as a function of time is a measurement of the cooling rate at the outer surface of the casting. The values of $h_{mold/casting}$ versus time are plotted and fitted using exponential functions, Figure 6. The exponential curve fittings functions are listed in Table I. The curve fitting exponent is used to measure for the cooling rate. If $h(t)_{mold/casting}$ has a higher exponent term, that would mean a faster cooling rate at the casting/mold interface.

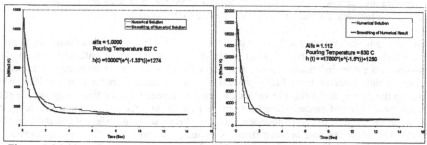

Figure 6. Mold-casting heat transfer function for Casting without vibration (left), and with vibration at 100 Hz and 49µm (right).

119

Table I. Mold-casting interface heat transfer coefficient functions

Vibration Condition (Hz, μm)	$h(t)_{mold/casting}$ (W/m² K)	Vibration Condition (Hz, μm)	$h(t)_{mold/casting}$ (W/m² K)
0,0	$10000(e^{(-1.35t)})+1274$	500, 6	$6000 (e^{(-1.3t)})+1280$
100, 49	$17000 (e^{(-1.5t)})+1250$	500, 12	$6000 (e^{(-0.9t)})+1100$
100, 99	$15000 (e^{(-1.4t)})+1100$	500, 24	$11000 (e^{(-1.65t)})+1100$
100, 149	$8000 (e^{(-1.35t)})+1300$	2000, 0.56	$18000 (e^{(-2t)})+1100$
100, 199	$12000 (e^{(-1.7t)})+1200$	2000, 1.1	$11000 (e^{(-2.8t)})+1000$
500, 3	$15000 (e^{(-1.9t)})+1000$	2000, 2.2	$17000 (e^{(-1.9t)})+1000$

Vibration at 100 Hz. Figure 6 shows that vibrating the solidified casting at 100 Hz will cause an increase in the cooling rate. This is seen from the higher exponent term seen in the $h(t)_{mold/casting}$ compared to that of casting without vibration as seen in. The increase on the exponent term was not seen for all cases of vibrating at 100 Hz but has an exception at 149μm amplitude, where the exponent is the same of as the unvibrated casting. This shows that at this level of frequency and amplitude, there are a different mechanism to refine the microstructure of the Al-12.6%Si other than increasing the cooling rate at the casting/mold interface.

Vibration at 500 Hz. When the casting is vibrated at a frequency of 500 Hz, the exponent term in $h(t)_{mold/casting}$ was not consistent. At 3μm the exponent term is 1.9, higher than 1.35 for the unvibrated case, which means the cooling at the casting/mold interface has improved. However, this is not the case when the amplitude increases to 6μm. The exponent term in this case has dropped down to 1.3 leading to the conclusion that cooling at the casting/mold interface is almost the same as that of the unvibrated case. Increasing the amplitude to 12μm, leads to a further reduction in the cooling rate at the casting/metal interface seen in the small value of the exponent term of the h(t) to 0.9. In spite of that, refinement of the lamellar spacing was seen in the center of the casting [2]. The cooling rate at the casting/mold interface is increased back a gain when the highest amplitude of 24μm as the exponent term is approximated to be 1.65.

Vibration at 2000 Hz. The cooling rate at the casting/mold interface is significantly increased when the frequency is increased to 2000Hz. The exponent term of the $h(t)_{mold/casting}$ is much higher than those predicted for casting without vibration and casting with lower frequencies. The combination of high frequency and very low amplitude is a key factor for significant improvement in the heat transfer coefficient significantly.

Phase Change. Figure 7 shows how phase-change is progressing with time for different cases of casting in the presence or absence of vibration. The zero time in the Figure 7 is the time where the temperature reaches 572°C. It has been observed that the solid does not start to appear upon reaching this temperature, but it takes some time due to the remelting process.

Two important observations can be seen in this Figure. The first one is that nucleation starts to initiate faster in casting where vibration is applied during than those without vibration. This observation is a clear indication that vibration plays an important role in assessing the nucleation process, which leads to the refining process. Vibration affected nucleation the least when vibrating the casting at 500Hz. The second observation comes from the time range for the solid fraction trend. The difference in time between the points where the solid fraction start to appear ($f_s > 0$) to the point of full solid development ($f_s = 1$) represents the effect of vibration on solid growth. The larger the time between these two points, the more time the eutectic constituents spend in growing and thus less refinement and vice versa.

120

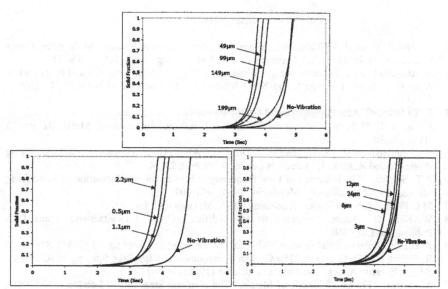

Figure 7. Solid fraction developments for casting with vibration
at different amplitudes and 100 Hz (top), at 500 Hz (bottom left) and 2kHz(bottom right).

Summary and Conclusion

This work used a simple 1-D numerical method to mimic the solidification of casting under various mold-vibration conditions. The developed temperature profiles and their proximity to the readings of the thermocouples prove that the model provides satisfactory accuracy. In addition, the simulated solidification time is comparable to the measured values. The results show that mold-vibration plays a major role in influencing the behavior of the metals as they solidify.

The significance of this work is that it shows the role that vibration plays in the nucleation and development of the solid phase. It was seen that the solid phase start to appear faster in casting solidified with the presence of vibration than the one solidified in a static mold. This result certainly supports the assumption that says that vibration aid the nucleation of the solidified metals. It is presumed that since the solid starts to grow faster, which means more nucleation instances are present. This would lead to a finer structure as proved in a previous work [2].

In most cases, the solid growth time is reduced as vibration is applied. The less the time for the solid fraction to grow, the finer the structure will be. However, the final structure should be decided from the combined effect of both the nucleation and solid growth. Therefore, it is fair to say that vibration effects on the microstructure is a result of the increased nucleation, faster cooling, and in some cases the reduction of solid growth time.

Acknowledgement

The support of King Fahd University of Petroleum and Minerals is acknowledged. The authors would also like to thank Secat Inc. for their support.

References

1. N. Abu-Dheir et al., "Silicon Morphology Modification in the Eutectic Al–Si Alloy Using Mechanical Mold Vibration," Materials Science Engineering, A393 (2005), 109–117.
2. N. Abu-Dheir et al., Towards Designing and Controlling the Microstructure and Properties of Aluminum Alloys Castings Using Mold Vibration," Transactions of NAMRI, 33 (2005), 311-317.
3. S. Ya Sokoloff, Acta Physicochem, 3 (1935), 939- 944.
4. J. Campbell, "Effects of Vibration During Solidification," International Metals Review, 2 (1981) 71-108.
5. V.O. Abramov et al., "Solidification of Aluminum Alloys Under Ultrasonic Irradiation Using Water-Cooled Resonator," Materials Letters, 37 (1998), 27-34.
6. R.T. Southin, "The Influence of Low Frequency Vibration on the Nucleation of Solidifying Metals," Journal Institute for Metals, 94 (1966), 401-407.
7. M.C Flemings, Solidification Processing (USA: McGraw-Hill Inc, 1974).
8. W. Kurz, D.J. Fisher, Fundamentals of Solidification, vol. 4 (Switzerland: Trans Tech Publications Ltd, 1998).
9. R. Elliott, "Eutectic Solidification," Materials Science and Engineering, 65 (1984), 85-92.
10. R. Elliott, S.M.D. Glenister, "The Growth Temperature and Inteflake Spacing in Aluminum Silicon Eutectic Alloys," Acta Metallurgica, 28 (1980), 1489-1494.
11. M.G. Day, "The Solidification of Metals," The Iron and Steel Ins. of London, 110 (1968), 177-183.
12. B. Tolouli, A. Hellawell, "Phase Separation and Undercooling in Al-Si Eutectic Alloy- The Influence of Freezing Rate and Temperature Gradient," Acta Metallurgica, 24 (1976), 565-573.
13. D.M. Stefanescu, "Methodologies for Modeling of Solidification Microstructure and Their Capabilities," The Iron and Steel Institute of Japan, 35 (1995), 700-707.
14. C.A. Santos, J.M.V. Quaresma, A. Garcia, "Determination of Transient Interfacial Heat Transfer Coefficients in Chill Mold Castings," Journal of Alloys and Compounds, 319 (2001), 174–186.
15. M.N. Srinivansan, "Heat Transfer Coefficient at the Casting-Mould Interface During Solidification of Flake Graphite Cast Iron in Metallic Moulds," Indian Journal of Technology, 20 (1982), 123-129.
16. N.J. Goudie, S. A. Argyropoulos, "Techniques for the Estimation of Thermal Resistance at Solid Metal Interfaces Formed During Solidification and Melting," Canadian Metallurgical Quarterly 34 (1995), 73-84.
17. M. Prates, H. Biloni, "Heat Flow Parameters Affecting Flow Parameters The Unidirectional Solidification of Pure Metals," Metallurgical Transactions, 3 (1972), 1501-1510.
18. M. Rappaz, "Modeling of Microstructure Formation in Solidification Processes," International Materials Review, 34 (1989), 93-123.
19. T.W. Clyne, "Modeling of Heat Flown in Solidification," Materials Science Engineering, 65 (1984), 111-124.
20. Metals Handbook vol.1, Ed 10, Properties and Selection: Irons, Steels, and High-Performance Alloys (ASM International, 1990).
21. M.D. Peres, C. A. Siqueira, A. Garcia, "Macrostructural and Microstructural Development in Al–Si alloys Directionally Solidified under Unsteady-State Conditions," Journal of Alloys and Compounds, 381 (2004), 168-181.
22. J. Sengupta et al., "Quantification of temperature, stress, and strain fields during the start-up phase of direct chill casting process by using a 3D fully coupled thermal and stress model for AA5182 ingots," Materials Science and Engineering A, 397 (1-2) (2005), 157–177.
23. J.V. Beck, B. Blackwell, C. R. ST. Clair, Inverse Heat Conduction (New York: Wiley, 1985).

Materials Processing under the Influence of External Fields
Edited by Qingyou Han, Gerard Ludtka, Qijie Zhai
TMS (The Minerals, Metals & Materials Society), 2007

DEGASSING OF MOLTEN ALUMINUM USING ULTRASONIC VIBRATIONS

Hanbing Xu[1], Thomas T. Meek[1], Qijie Zhai[2], and Qingyou Han[3]

[1]The University of Tennessee, 434 Dougherty Hall, Knoxville, TN 37996, USA
[2]Shanghai University, P.R. China
[3]Oak Ridge National Laboratory, Oak Ridge, TN 37831, USA

Keywords: Ultrasonic Vibration, Degassing, Hydrogen Content, and Aluminum Alloys

Abstract

In order to investigate the effects of ultrasonic vibration on degassing of aluminum alloys, three experimental systems have been designed and built: one for ultrasonic degassing in open air, one for ultrasonic degassing under reduced pressure, and one for ultrasonic degassing with a purging gas. Experiments were first carried out in air to test degassing using ultrasonic vibration alone. The limitations with ultrasonic degassing were outlined. Further experiments were then performed under reduced pressures and in combination with purging argon gas. Experimental results suggest that ultrasonic vibration alone is efficient for degassing a small volume of melt. Ultrasonic vibration can be used for assisting vacuum degassing, making vacuum degassing much faster than that without using ultrasonic vibration. Ultrasonically assisted argon degassing is the fastest method for degassing among the three methods tested in this research. More importantly, dross formation during ultrasonically assisted argon degassing is much less than that during argon degassing. The mechanisms of ultrasonic degassing are discussed.

Introduction

Porosity is one of the major defects in aluminum alloy shape castings. Porosity in castings occurs because gas participates from solution during solidification or because the liquid metal cannot feed through the interdendritic regions to compensate for the volume shrinkage associated with solidification. Hydrogen is the only gas that is appreciably soluble in molten aluminum [1, 2]. The removal of the dissolved hydrogen in the molten aluminum alloy is critical for the production of high-quality castings.

Several methods are currently in use to degas aluminum [3-5]. These methods include rotary degassing using nitrogen or argon or mixture of either of these with chlorine as a purge gas [3, 4], tablet degassing using hexachloroethane (C_2Cl_6) [3, 4], vacuum degassing [6-8], and ultrasonic degassing [9-13]. Ultrasonic degassing, an environmentally clean and relative inexpensive technique, uses high intensity ultrasonic vibrations to generate oscillating pressures in molten aluminum. Degassing requires the introduction of acoustic energy in the melt of a sufficient intensity to set up a pressure variation that will initiate cavitation [10, 14]. Hydrogen is removed by diffusing to the cavitation bubbles and escaping from the melt with the bubbles. The purpose of this article is to assess the effectiveness of ultrasonic degassing and to explore the possibility of using ultrasonic vibration to assist vacuum degassing and purging gas degassing.

Experimental Methods

Three experimental systems have been designed and built: one for ultrasonic degassing alone, one for ultrasonic degassing under reduced pressure, and one for ultrasonic degassing with a purging gas. The composition of the aluminum alloy is listed in Table 1. Reduced pressure test (RPT) was used to determine the porosity level of the melt. The porosity level thus measured was then correlated to the hydrogen concentration using a calibration curve obtained using the Leco™ hydrogen analysis. Details of the hydrogen measurement are reported in another article [15].

Table 1: Chemical Composition of A356 alloy

Element	Al	Cu	Fe	Mg	Mn	Si	Ti	Zn
Wt.%	92.5	0.1	0.1	0.35	0.05	7.2	0.1	0.05

Ultrasonic Degassing: The experimental system for ultrasonic degassing under normal atmospheric pressure consisted of a 20 kHz ultrasonic generator, an air-cooled converter made of piezoelectric lead zirconate titanate (PZT) crystals, a booster, a probe, an acoustic radiator to transmit ultrasonic vibration into aluminum melt, and a furnace in which the aluminum melt was held. The transducer was capable of converting up to 1.5 kW of electric energy at a resonant frequency of 20 kHz. The amplitude of the ultrasonic vibration could be continuously adjusted from 30% to 100% of 81 μm, which is the maximum amplitude of the unit. During experiments, ultrasound was injected into the aluminum melt by using a cylindrical radiator made of titanium alloy Ti-6Al-4V. The aluminum alloy was held in a graphite crucible and melted in the electric furnace. The temperature of the melt was controlled within an accuracy of ±10 °C. After the melt was heated to a predetermined temperature, a preheated ultrasonic radiator was inserted in molten metal. Ultrasonic vibration was then applied in the molten metal for certain amount of time before samples were taken for hydrogen measurements.

Ultrasonic Degassing under Reduced Pressure: A steel vacuum chamber was built for vacuum degassing with the assistance of ultrasonic vibrations. The crucible inside the electric furnace could hold molten aluminum alloys weighing up to 800 g. The minimum remnant pressure of this vacuum chamber was 50 mTorr. It took a few seconds for the vacuum chamber to reach 100 Torr and 10 Torr, around 1 minute to reach 1 Torr, and 20 to 30 minutes to reach 0.1 Torr.

Ultrasonic Degassing Combined with Argon Degassing: The system is similar to the used for ultrasonic degassing in open air except that an acoustic probe with gas purging capability was fabricated. Argon was introduced from the center of the probe to the molten metal. A flowmeter was used to control and monitor the flow rate of argon. Ultrasonic vibration was injected from the top of the melt.

Experimental Results

Ultrasonic Degassing

Ultrasonic degassing was carried out for an aluminum A356 melt under three conditions. These conditions included the humidity of the air, the temperature of the melt, and the volume/size of the melt. The humidity was varied from 40% to 60%. Four melt temperatures, 620°C, 660°C, 700°C and 740°C were tested. The weight of the melt was 0.2 kg, 0.6 kg and 2 kg, respectively. Figure 1 shows the ultrasonic degassing rates in molten A356 alloy prepared at 740°C at 40 and 60% humidity or with differing initial hydrogen concentrations. The experiments were carried out using a crucible containing 0.2 kg of aluminum melt. Without ultrasonic vibration, the hydrogen content of the specimen cast under a humidity of 60% was much higher than that cast under a humidity of 40%. The initial hydrogen concentration was 0.45 and 0.36 ppm at humidity levels of 40 and 60% respectively. With ultrasonic vibrations, the hydrogen contents decreased sharply with increasing ultrasonic processing time in the first minute and then reached a plateau density, which corresponds to the steady-state hydrogen concentration in the melt at 740°C. This trend was true for specimens cast under both humidity levels. The results shown in Figure 1 suggest that degassing in a small aluminum melt was extremely fast. No matter what the initial hydrogen concentrations were, degassing can be achieved within one minute. The humidity has little effect on the time required for degassing using ultrasonic vibrations. The hydrogen content at the plateau density shown in Figure 1 was 0.14 ppm.

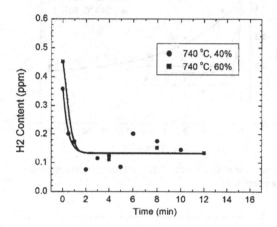

Figure 1: The hydrogen content as a function of ultrasonic processing time in melts prepared at different humidity.

Figure 2 shows the efficiency of ultrasonic degassing in A356 alloy melts under various melt temperatures. The results were obtained in a crucible containing 0.2 kg aluminum alloy. As illustrated in Figure 2, it took about one minute of ultrasonic vibration for the melt to reach a steady-state hydrogen content plateau when the melt was ultrasonically processed at a temperature of 700°C or 740°. The processing time required to degas the melt (to reach the steady-state hydrogen content plateau) increased with decreasing melt temperature in the temperature of 620 to 700 °C. It took almost 10 minutes to degas the melt held at 620°C, much longer than in the melt held at temperatures higher than 700°C.

The data plotted in Figure 2 also indicates that the plateau hydrogen content is not sensitive to the processing temperature in the range between 620°C and 740°C. It takes longer processing time to reach the plateau hydrogen content when the processing temperature is low but once the plateau hydrogen content is reached, the porosity levels in the specimens are identical. The plateau hydrogen concentration is in the range of 0.1 to 0.2 ppm.

In order to obtain a quantitative evaluation of the degassing speed in molten aluminum, the weight (volume) of the melt was varied. The experiments were carried out in melt at 700°C under a humidity of 60%. The weights of the melts were 0.2 kg, 0.6 kg and 2.0 kg, respectively. As illustrated in Figure 3, the ultrasonic processing time required for reaching the steady-state plateau hydrogen content increases with increasing weight (or volume) of the melt. Degassing times were as follows: 0.2-kg melt — 1 min; 0.6-kg melt — 4 min; 2.0-kg melt — almost 7 min. Degassing speed in a larger volume melt is much slower than that in a smaller melt. However it is still encouraging that the steady-state plateau hydrogen content is not sensitive to the volume of the melt. Fast degassing of a large-volume melt can be obtained by use of multiple ultrasonic radiators to inject ultrasonic vibrations into the melt.

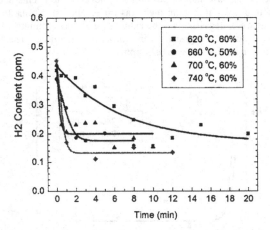

Figure 2: The hydrogen content as a function of ultrasonic processing time in melt degassed at different processing temperatures.

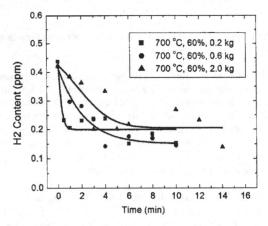

Figure 3: The hydrogen content as a function of ultrasonic processing time in melt of different sizes.

Ultrasonic Degassing under Reduced Pressures

Ultrasonic degassing under reduced pressure was also evaluated. The vacuum degassing data were used as baseline data for evaluating ultrasonic degassing under reduced pressures. Vacuum degassing was carried out in 0.6 -kg aluminum samples. The process parameters studied under vacuum degassing included the remnant pressure and degassing time. The remnant pressure was varied from 760 torr to 0.1 torr and the degassing time was varied from 1 to 45 minutes. For the remnant pressure of 100 torr and 10 torr, the experiments were conducted at temperature of 720°C with a humidity of ~60%. For remnant pressure was 0.1 and 1 torr, the humidity was maintained at ~50%. Vacuum degassing with the assistance of ultrasonic vibration was performed under two different remnant pressures: 100 torr and 1 torr, respectively. The experiments were carried out with 0.6 -kg melts at 720°C with a humidity of 50%. It took several seconds for the remnant pressure to reach 100 torr and around 1 minute to reach 1 torr.

The hydrogen contents as a function of processing time under various conditions are shown in Figure 4. Data points marked by filled squares indicate the efficiency of ultrasonic degassing under two remnant pressure levels (100 and 1 torr). As the figure indicates, the plateau hydrogen content was attained much faster through the use of the combination of ultrasonic degassing and vacuum degassing than by use of either ultrasonic degassing or vacuum degassing alone. For comparison, Figure 4 also display data obtained using ultrasonic vibration under normal pressure (in air) (marked on the figures with triangles) and data obtained using vacuum degassing (marked with filled circles). As is illustrated by these two figures, the most efficient degassing method is ultrasonic degassing under reduced pressure and the slowest is vacuum degassing.

The processing times required to reach the steady-state plateau hydrogen content for the three methods is also shown in Figure 4. It took just 1 minute for ultrasonic degassing under reduced pressure to reach the steady-state density, 4 minutes for ultrasonic degassing alone, and more than 20 minutes for vacuum degassing. Even under a partial vacuum condition such as 100 Torr, the efficiency of ultrasonic degassing could be increased by using ultrasonic degassing under a reduced pressure.

Ultrasonically Assisted Argon Degassing

Due to their inherent limitations, methods such as ultrasonic degassing and ultrasonic degassing under reduced pressure cannot be used for a fast degassing of a large volume aluminum melt. This is partly because there is a large attenuation of ultrasonic vibration in liquids. The intensity of ultrasonic vibrations decreases sharply with increasing distance from the ultrasonic radiator. On the other hand, degassing with argon is also a relatively slow process due to the large size of argon bubbles. If, howerer, ultrasonic vibration is applied to the melt, the pressure induced by acoustic waves could break up the large argon bubbles into numerous smaller bubbles. As a result, the area of the bubble surface in the melt will be increased substantially. This should be favorable for more efficient degassing. Based on such considerations, ultrasonically assisted argon degassing was tested.

Figure 5 shows the hydrogen content of the melt as a function of processing time in 5kg melts processed using either argon degassing or ultrasonically assisted argon degassing. As illustrated in Figure 5, ultrasonically assisted argon degassing is more efficient for degassing than argon degassing. When ultrasonic assisted argon degassing was used, the aluminum melt reached the steady-state plateau hydrogen content in 5 minutes, whereas, more than 10 minutes were required when argon degassing alone was used. The efficiency of ultrasonically assisted argon degassing is almost two times that of argon degassing.

Despite the much greater efficiency of the ultrasonically assisted argon degassing technique, in our experiments there were still limitations to this method. Because the ultrasonic vibration was injected from the top of the melt, the bubbles released from the ultrasonic radiator could not penetrate deep into the melt due to the large viscous drag force on the bubbles in molten aluminum. As a result, the volume of the molten metal that the argon bubble passed through was limited. An increase in the vibration amplitude or the argon flow rate can increase the volume of molten aluminum being degassed.

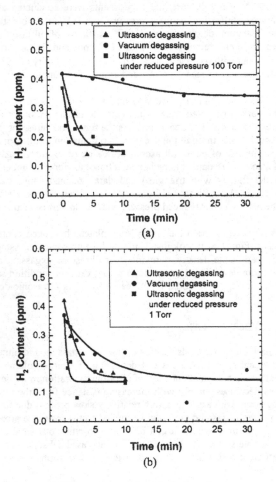

Figure 4: The efficiency of different degassing techniques when remnant pressure is (a) 100 torr and (b) 1 torr. The degassing techniques include ultrasonic degassing, vacuum degassing and ultrasonic degassing under reduced pressure.

The benefit of argon degassing with ultrasonic vibration also includes a reduction of dross formation during the degassing process. When the large argon bubbles were injected into the molten metal, the top surface of the melt was turbulent as the large bubbles escaping from the melt. The oxide layer on the melt top surface was broken so the molten aluminum was exposed to the air, forming more oxides. The use of ultrasonic vibrations broken up the large argon bubbles into tiny bubbles and made the melt surface less turbulent. Furthermore, the shorter degassing time with the use of ultrasonic processing meant that much less dross was formed.

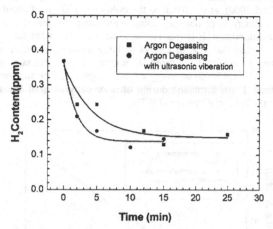

Time (min)

Figure 5: A comparison of hydrogen content as a function of processing time for argon degassing and ultrasonically assisted argon degassing in a 5 kg melt.

Figure 6 shows the amount of dross formation during argon degassing and ultrasonically assisted argon degassing. The line at the top of the graph illustrates the weight of dross formed during argon degassing of a 5 -kg aluminum melt. The bottom line was obtained using ultrasonically assisted argon degassing. Table 2 lists the measured weight of dross formed during degassing using argon degassing without and with ultrasonic processing. At a processing time of 15 min, the amount of dross formed with argon degassing was 39.4 g but only 6.4 g with ultrasonically assisted degassing. Thus, six times more dross formed when argon degassing was used without the ultrasonic processing. And taking into consideration the much shorter degassing with the ultrasonically assisted argon degassing method (5 minutes vs 10 min), dross formation during ultrasonically assisted argon degassing was much less than one-sixth of that produced during argon degassing. As can be seen in Table 2, for instance, during a 5-min ultrasonically assisted argon degassing, dross formation was only 1.8 g; by contrast, during a 10-min argon degassing, dross formation was 22.8 g. This means that the use of ultrasonically assisted degassing can achieve a reduction of about 92% of dross formation over the use of argon degassing method in degassing of molten aluminum alloys.

Table 2: Amount (grams) of dross formed at various degassing times using argon degassing and ultrasonically assisted argon degassing.

Degassing method	Degassing time (min)					
	2	5	10	15	20	25
Argon degassing alone		16.6	*22.8*[a]	39.4	49.5	65.1
Argon degassing with ultrasonics	0.9	*1.8*[a]	5.2	6.4	NA	NA

[a] Bold italic indicates the point at which degassing was completed.

It should be noted that our experimental data on dross formation during argon degassing were higher than the industry average. Using out experimental data, if one assumes that argon degassing of 5 kg of molten aluminum has been completed in 10 min, the percentage of dross formation is 0.46% (22.8/5000), as compared to the industrial average of about 0.2%. However the average data for industry were obtained using equipment capable of processing 60,000–140,000 lb/h of molten aluminum. It is reasonable to assume that dross formation in a small-volume melt is much higher than that in a large volume during a turbulent argon degassing process. It is interesting to note that the dross formation during ultrasonically assisted argon degassing was only 0.036% (1.8/5000) assuming a 5-min degassing time and was 0.1% assuming a 10-min degassing time. Dross formation during ultrasonically assisted degassing of a small melt is much less than the industrial average of 0.2%.

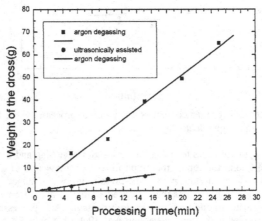

Figure 6: The amount of dross formation as a function of processing time during argon degassing and ultrasonically assisted argon degassing.

Mechanisms of Ultrasonic Degassing

The injection of ultrasonic vibrations in molten aluminum causes alternating pressure in the melt. While pressure can affect the solubility of hydrogen in molten aluminum, alternating pressure especially pressure that is alternating symmetrically at high frequencies, cannot affect solubility. However, a sound wave with pressure above the cavitation threshold propagating through a liquid metal generates cavitation bubbles in the melt and drastically alters the mass transfer of gas from the solution to the bubbles.

Cavitation consists of the formation of tiny discontinuities or cavities in liquids, followed by their growth, pulsation, and collapse. Cavities appear as a result of the tensile stress produced by an acoustic wave in the rarefaction phase. If the tensile stress (or negative pressure) persists after the cavity has been formed, the cavity will expand to several times the initial size (During cavitation in an ultrasonic field, many cavities appear simultaneously at distances less than the ultrasonic wavelength). In this case, the cavity bubbles retain their spherical form. The subsequent behavior of the cavitation bubbles is highly variable: a small fraction of the bubbles coalesces to form large bubbles but most are collapsed by an acoustic wave in the compression phase. In a melt containing dissolved gases, the gaseous elements diffuse into the cavitation bubbles through the bubble/melt interfaces during the nucleation and growth stage of the cavitation bubbles. Some of the dissolved gases escape when cavitation bubbles at the molten surface collapse, thereby resulting in degassing.

Generally ultrasonic degassing can be divided into three stages: (1) nucleation of cavitation bubbles on nuclei (usually solid inclusions containing cavities) and growth of the bubbles due to the diffusion of hydrogen atoms from the surrounding melt to the bubbles, (2) coalescence of bubbles to form large bubbles, and (3) float of large bubbles to the surface of the molten metal and escape of the bubbles at the top melt surface.

The first stage of ultrasonic degassing, nucleation and growth of cavitation bubbles, can be easily achieved using high intensity ultrasonic vibration. The ultrasonic unit used in this project has the power density much higher than the critical power density for creating cavitation. As a result, cavitation can be readily achieved when the unit is operated at amplitudes higher than 30% of the maximum amplitude. Due to the high attenuation of ultrasonic vibration in the liquid, cavitation mainly forms near the ultrasonic radiator and the cavitation bubbles are transported to the bulk liquid by ultrasonic streaming. Since the cavitation bubbles are small (in the range of microns) and the interfacial area of the bubble/melt interfaces is large, degassing is extremely fast if most of the cavitation bubbles escape from the melt. This is what happens when ultrasonic vibration is used for degassing a small volume of melt. Due to flow produced by acoustic streaming, the cavitation bubbles formed near the ultrasonic radiator are quickly transported to the melt surface, bringing with them the dissolved gases, which then escape the melt. Use of the reduced pressure of a vacuum facilitates the escape of the gasses at the melt surfaces. This accounts for the facts that ultrasonic degassing and the ultrasonic degassing under a reduced pressure are two of the methods which enable of fast degassing of a small volume melt.

One critical issue of ultrasonic degassing is the survival of the cavitation bubbles in the melt which affects its degassing efficiency. During the initial stages of the nucleation and growth of the cavitation bubbles, the dissolved gases diffuse into the bubble and form gas molecules. As these bubbles travel away from the ultrasonic radiator or as the tensile stress become less tensile or even compressive, the gas molecules can decompose at the bubble/melt interfaces and dissolve back into the melt. The gases in the cavitation bubbles will also dissolve into the melt when the cavitation bubbles collapse. In fact, most of the cavitation bubbles cannot survive longer enough to reach the melt surface. Based on the understanding of the mechanism of ultrasonic degassing, a novel approach for ultrasonic degassing has been developed. The new approach involves the use of a small amount of purging gas (argon) for ultrasonic degassing. The idea is to use this small amount of purging gas to extend the survival time of the cavitation bubbles which contain dissolved gas as the bubbles nucleate and grow. In practice, the purging gas is introduced through the ultrasonic radiator. High intensity ultrasonic vibration is fed through the radiator to break up the purging gas bubbles as well as to create a large amount of cavitation bubbles. The small purging gas bubbles can survive in the melt forever because argon will not dissolve into the melt. As the purging gas bubbles travel from the radiator to the bulk melt, they collect the cavitation bubbles which contains dissolved gases. On the other hand, acoustic streaming helps to distribute the small purging gas bubbles uniformly throughout the melt, improving the degassing efficiency further. The experimental results show that the new degassing method is more efficient than the traditional purging gas methods.

Conclusions

Ultrasonic degassing is an efficient way of degassing a small volume melt. The early stage of ultrasonic degassing is extremely fast. It takes only a few minutes to degas a small volume melt and to reach a steady-state plateau hydrogen content. The humidity/initial hydrogen concentration has little effect on the degassing efficiency using ultrasonic vibrations. The melt temperature has a significant effect on the efficiency of ultrasonic degassing. The degassing rate in the temperature range between 700°C and 740°C is faster than that in the temperature range between 620°C and 660°C. These results clearly suggest that ultrasonic vibration alone is efficient for degassing a small volume of melt. The applications would be degassing a shallow aluminum trough for transporting aluminum from a melting furnace to a casting machine. Often multiple radiators have to be used for degassing a large volume of molten aluminum.

The combination of ultrasonic degassing and vacuum degassing results in a much faster degassing than does the use of ultrasonic degassing alone. Even under a partial vacuum condition such as 100 torr, the efficiency of ultrasonic degassing could be increased by using ultrasonic degassing under a reduced pressure.

Ultrasonically assisted argon degassing is a more efficient method than argon degassing alone. For a 5 –kg aluminum melt, a steady–state hydrogen content plateau was reached in only 5 min using ultrasonically assisted degassing, compared to more than 10 min using argon degassing. More importantly, dross formation during ultrasonically assisted argon degassing is much less than that during argon degassing.

Acknowledgment

This research was supported by the United States Department of Energy, Assistant Secretary for Energy Efficiency and Renewable Energy, Industrial Technologies Program, Industrial Materials for the Future (IMF), Aluminum Industry of the Future, under contract No. DE-FC36-02ID14399 with UT-Battelle, LLC. The authors would like to thank Ohio Valley Aluminum Co. and Secat Inc for providing industrial support.

References

1.   D.R. Poirier, K. Yeum, and A.L. Maples, Metall. Trans., 18A (1987) 1979-1987.
2.   Q. Han, and S. Viswanathan, Metall. Mater. Trans., 33A (2002), 2067-2072.
3.   A. M. Samuel and F.H. Samuel, Journal of Materials Science, 27(1992), 6533-6563.
9th Edition, Metals Handbook, Volume 15, ASM International, 1988.
4.   L. W. Eastwood, Gas in Light Alloy, John Wiley and Sons, Inc., New York, 1946
5.   O. Richly, Aluminium. 57(8) (1981) 546-548.
6.   A. Choudhury, M. Lorke, Aluminium. 65(5) (1989) 462-468.
7.   G. Popescu, I. Gheorghe, Materials Science Forum, 147, (1996) 217-222.
8.   O. V. Abramov, Ultrasound in Liquid and Solid Metals, CRC Press, London, 1994.
9.   G. I. Eskin, Ultrasonic Treatment of Light Alloy Melts, Gordon & Breach, Amsterdam, 1998.
10.   I. G. Brodova, P. S. Popel and G. I. Eskin, Liquid Metal Processing, Taylor & Francis Inc, New York, 2002.
11.   G. I. Eskin, Advanced Performance Materials, 4(1997), 223-232.
12.   G. I. Eskin, Metallurgist, 42(8) (1998) 284-291.
O. G. Abramov, High-Intensity Ultrasonics, Gordon and Breach Science Publishers, Amsterdam, The Netherlands, 1998.
13.   L. D. Rozenberg (ed.), Physical Principles of Ultrasonic Technology – vol. 2, Plenum Press, New York, 1973.
14.   L.D. Rozenberg (ed.), Sources of High-Intensity Ultrasound, Vol. 1 and 2, Plenum Press, New York, 1969.
15.   H. Xu, T.T. Meek, and Q. Han, Effect of Ultrasonic Vibration on Degassing of Aluminum Alloys, submitted to Materials Science and Engineering A.

Materials Processing under the Influence of External Fields
Edited by Qingyou Han, Gerard Ludtka, Qijie Zhai
TMS (The Minerals, Metals & Materials Society), 2007

RESEARCH PROGRESS ON ULTRASONIC CAVITATION BASED DISPERSION OF NANOPARTICLES IN AL/MG MELTS FOR SOLIDIFICATION PROCESSING OF BULK LIGHTWEIGHT METAL MATRIX NANOCOMPOSITE

Xiaochun Li

Department of Mechanical Engineering
University of Wisconsin-Madison, Madison, WI53706

Keywords: Ultrasonic cavitation, Solidification Processing, Metal matrix nanocomposites, Lightweight

Abstract

Aluminum and magnesium structural components are of significance for numerous applications. The properties of aluminum/magnesium alloys would be enhanced considerably if reinforced by ceramic nanoparticles. However, it is very challenging to disperse the nanoparticles uniformly in the metal melts for solidification processing of bulk nanocomposites. In this study, a novel method, ultrasonic cavitation based dispersion of nanoparticles in aluminum A356 and Mg melts for solidification processing (e.g. casting) is experimented. With only 1.0wt.% nano-sized SiC reinforcement, the ultimate tensile strength and yield strength of the aluminum alloy A356 were enhanced approximately 100% while the ductility was retained. Meanwhile, about 40% reinforcement on UTS and 60% enhancement on yield strength were achieved on Mg in our preliminary study. The study on micro/nano structures of the nanocomposites validates that a uniform distribution and effective dispersion of nanoparticles in the matrix were achieved. This study paves a way for high volume processing of Al/Mg nanocompoistes for industrial applications.

Introduction

In recent years, significant' effort has been taken to develop metal matrix nanocomposites (MMNCs) [1-12]. Compared with conventional metal matrix composites (MMCs) that are reinforced with micro inclusions, this new class of material can overcome many disadvantages in MMCs [13-15], such as poor ductility, low fracture toughness and machinability. The properties of metals would be enhanced considerably if reinforced by ceramic nanoparticles (less than 100nm). Moreover, MMNCs could offer a significantly improved performance at elevated temperatures [7-10]. MMNCs are poised to generate enormous impact on aerospace, automobile, and military industries.

Although MMNCs have a great potential for widespread applications, the current processing technologies are neither reliable nor cost effective to enable a high volume and net shape production of complex MMNC structural components with reproducible structures and properties. Traditional fabrication methods, such as high energy ball milling, rapid solidification, electroplating, and sputtering etc, can not be used for mass production and net shape fabrication of complex structural components without significant post processing [7-12].

In this research, a new method that combines solidification processes with ultrasonic cavitation based dispersion of nanoparticles in metal melts has been developed. Ultrasonic cavitation can produce transient (in the order of nanoseconds) micro "hot spots" that can have temperatures of about 5000°C, pressures above 1000 atms, and heating and cooling rates above 10^{10} °K/s [16]. The strong impact coupling with local high temperatures can potentially break nanoparticle clusters and clean the particle surface. Since nanoparticle clusters are loosely packed together, air could be trapped inside the voids in the clusters, which will serve as nuclei for cavitations. The size of the clusters ranges from nano to micro due to the attraction force among nanoparticles and the poor wettability between the nano particles and metal melt.

Material System Selection

Aluminum alloy A356 was selected as the matrix material for Al matrix nanocomposites since it is also readily cast-able and widely studied. The chemical composition of the A356 alloy is shown in Table 1.

Table 1 Nominal chemical composition of matrix alloy A356 (wt%)

Si	Fe	Cu	Mn	Mg	Zn	Ti	Al
6.5-7.5	0.20	0.20	0.10	0.25-0.45	0.10	0.20	bal

Pure magnesium (99.8%) was selected as the matrix to produce magnesium composites.
The nano-sized ceramic particles used in this study were β-SiC (spherical shape, average diameter<= 30nm, Composition: SiC>=95%, [O]:1~1.5%, [C]: 1~2%).

Experimental Setup and Procedure

As shown in Fig.1, the experimental setup consisted of processing and controlling parts. An electric resistance heating unit was used to melt the A356 in a small graphite crucible with a casting capacity of approximately 2 lbs, controlled with an electric transformer. A Permendur power ultrasonic probe with Niobium waveguide, which was coupled to a 17.5 kHz and maximum 4.0 kW power output (Advanced Sonics, LLC, Oxford, CT), was dipped into the melt for ultrasonic processing. The melt was protected with four Argon shield gas nozzles for A356 or by CO_2/SF_6 (volume ratio 99:1) mixed gas for Mg. The processing temperature was monitored with a steel sheath thermal couple probe. Nano-sized SiC particles were added into melts from the top of crucible during the process.

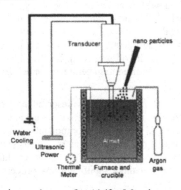

Fig. 1. Schematic of experimental setup for Al (for Mg, the protective gas is CO_2/SF_6)

To test the mechanical properties of the MMNCs, specimens with a diameter of 0.35 inch (8.89mm) and a gage length of 1.25 inch (31.75mm) were tested on a Sintech 10/GL with an extensometer.

Scanning Electron Microscopy (SEM) images were obtained with LEO1530. Transmission Electron Microscopy (TEM) Philips CM200 was used to study the nanostructure of MMNCs.

Results and Discussion

Aluminum Nanocomposites:
To optimize the processing parameters, design of experiment (DOE) method was used. It was determined that a processing time of 60 minutes was adequate to disperse 1.0wt.% nanoparticles in A356 melt and ultrasonic power and processing temperature were around 3.0 kW and 700 °C respectively. With those optimized parameters, and only 1.0 wt.% of SiC nanoparticles, the mechanical properties of nanocomposites are improved near 100% with little change in elongation.

Table 2: Comparison of mechanical properties between nanocomposite and A356

	A356 with 1.0wt.% SiC nanoparticle (after T6)	Original A356 (after T6)
Tensile Strength (ksi)	69.86	37.37
Yield Strength (ksi)	53.7	27.8
Elongation	6.65	6.67

Samples of A356 matrix and nanocomposite after T6 heat treatment were examined by SEM. Fig.2 clearly shows that nano-sized SiC particles were uniformly distributed and effectively dispersed in the A356 matrix. It should be noted that some micro clusters still remain in the matrix.

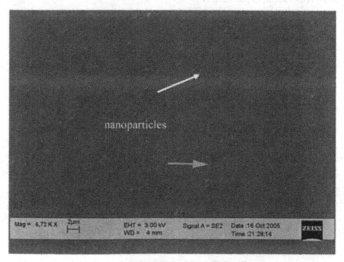

Fig. 2. SEM images of A356 nanocomposite

135

Fig. 3(a) shows a typical bright-field TEM image of the nanocomposite. The image clearly shows the single particles were dispersed in the matrix. In Fig.3b, the lattice of the Al matrix is clearly shown by High Resolution transmission electron microscopy (HRTEM). However it is very difficult to find a particle also oriented in a low index at the zone axis, making it impossible to reveal the lattices of the Al matrix and the nanoparticle simultaneously. Nevertheless, the image clearly shows that the lattice of Al gradually evolved to that of SiC, This gradual transition indicates no lattice distortions or mismatch, which could have been eliminated by the T6 treatment.

(a)   (b)

Fig. 3. (a) Bright Field TEM image of Al nanocomposite (b) bonding zone around single particle in Al matrix

Mg Nanocomposites:

Fig.4 shows the morphology and distribution of SiC nanoparticles in the Mg matrix. It can be seen that nanoparticles are well distributed and dispersed, although there are some SiC agglomerates ~100-300nm in the matrix.

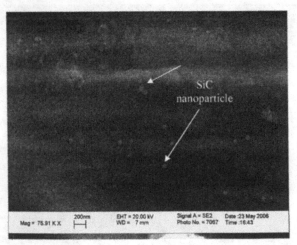

Fig 4. Pure Mg/2% SiC nanoparticles

Preliminary tensile tests were done on the Mg/2SiC composite samples. Significant increase in UTS and yield strength were achieved, while the elongation property of the composites varied. Preliminary tensile test result on Mg/2%SiC is shown in Fig.5. Over 60% increase on yield strength and 40% increase on UTS were achieved.

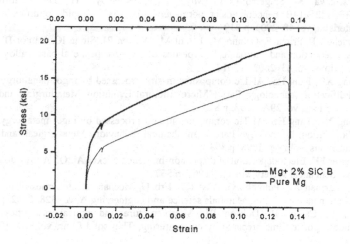

Fig.5. Preliminary tensile results on Mg/2%SiC composite

Conclusion

In this study, the feasibility of ultrasonic cavitation based dispersion of nanoparticles in Al and Mg melts was validated. An experimental system for ultrasonic cavitation based solidification processing was fully developed and Al alloy A356 and Mg nanocomposites were fabricated and characterized. With optimized processing parameters, the tensile test results showed that, with only 1.0wt% nano-sized SiC, the ultimate tensile strength (UTS) and Yield strength of the nanocomposites were improved approximately 100% while ductility is retained. Meanwhile, about 40% reinforcement on UTS and 60% enhancement on yield strength were achieved on Mg in our preliminary study. Micro/nano structure study shows that a good nanoparticle distribution and dispersion in the Al and Mg matrix were achieved. By use of HRTEM, the nanostructure study indicates that a coherent bonding was obtained and the lattice of the Al matrix gradually evolved to that of the nanoparticles.

Acknowledgements

This work is supported by the American Foundry Society and National Science Foundation.

References

[1]   Jiang QC, Li XL, Wang HY, Fabrication of TiC particulate reinforced magnesium matrix composites, Scripta Materialia, vol.48. 2003. p.713
[2]   Rawal S, Metal-matrix composites for space applications, JOM, vol.53. 2001. p.14

[3]   Crainic N, Marques AT, Nanocomposites: a state-of-Art Review, Key engineering materials, vol.230-232. 2002. p.656

[4]   Yamasaki T, Zheng YJ, Ogino Y, Terasawa M, Mitamura T, Fukami T, Formation of metal-TiN/TiC nanocomposite powders by mechanical alloying and their consolidation, Materials Science and Engineering A, vol.350A. 2003. p.168

[5]   Hirsosawa S, Shigemoto Y, Miyoshi T, Kanekiyo H, Direct formation of $Fe_3B/Nd_2Fe_{14}B$ nanocomposite permanent magnets in rapid solidification, Scripta materialia, vol.48. 2003. p.839

[6]   Audebert F, Prima F, Galano M, Tomut M, Warren PJ, Stone IC, Cantor B, Structural characterization and mechanical properties of nanocomposite al-based alloy, Materials transitions, vol.43. 2002. p.2017

[7]   Tong XC, Fang HS, Al-Tic composites in situ-processed by ingot metallurgy and rapid solidification technology: Part 1 Microstructural Evolution, Metallurgical and materials Transactions, vol.29A. 1998. p.875

[8]   Tong XC, Fang HS, Al-Tic composites in situ-processed by ingot metallurgy and rapid solidification technology: Part 2 mechanical behavior, Metallurgical and materials Transitions, vol.29A. 1998. p.893

[9]   Sautter FK, Electodeposition of dispersion-hardened Nickel-Al2O3 Alloys, Journal of the electrochemical society, vol.110. 1963. p.557

[10]   Zimmerman AF, Pulumbo G, Aust KT, Erb U, Mechanical properties of nickel silicon carbide nanocomposites, Materials science and engineering A, vol.328. 2002. p.137

[11]   Niu F, Cantor B, Dobson PJ, Microstructure and optical properties of Si-Ag nanocompoiste films prepared by co-sputtering, Thin solid films, vol.320. 1998. p.184-191

[12]   Islamgaliev RK, Yunusova NF, Sabirov IN, Sergueeva AV, Valiev RZ, Deformation behavior of nanostructured aluminum alloy processed by severe plastic deformation, Materials science and engineering A, vol.319. 2001. p.877

[13]   Arsenault RJ, Wang L, Feng CR, Strengthening of composites due to microstructural changes in the matrix, Acta metal. Mater. Vol.39. 1990. p.47

[14]   Nardone VC, Prewo KM, On the strength of discontinuous silicon carbide reinforced aluminum composites, Scripta metallugica, vol.20. 1986. p.43

[15] Kim Y-W, Griffith WM, Dispersion strengthened aluminum alloys, Publication of TMS, 1998; 27:220

[16]   Suslick S, Didenko Y, Fang MM, Hyeon T, Kolbeck KJ, McNamara WB, Midleleni MM, Wong M, Phil. Tans. R. Soc. Lond., vol.A357. 1999. p.335

Materials Processing under the Influence of External Fields
Edited by Qingyou Han, Gerard Ludtka, Qijie Zhai
TMS (The Minerals, Metals & Materials Society), 2007

EXPERIMENTAL INVESTIGATIONS ON ROTARY ULTRASONIC MACHINING OF HARD-TO-MACHINE MATERIALS

N.J. Churi [1], Z.J. Pei [1], C. Treadwell [2]

[1] Department of Industrial and Manufacturing Systems Engineering,
Kansas State University, Manhattan, KS 66506, USA
[2] Sonic-Mill, 7500 Bluewater Road, Albuquerque, NM 87121, USA

Keywords: Ceramic, chipping, composite, cutting force, grinding, rotary ultrasonic machining, surface roughness, titanium alloy.

Abstract

Rotary ultrasonic machining (RUM) is a hybrid process combining the material removal mechanisms of ultrasonic machining and diamond grinding. Its attractive features make it suitable for machining certain hard-to-machine materials. This paper reports experimental investigations on RUM of several hard-to-machine materials, including alumina, silicon carbide, ceramic-matrix composites, and titanium alloys. The advantage of RUM over diamond grinding is demonstrated by experiments with and without ultrasonic vibration of the cutting tool. The effects of process parameters on the RUM performance (in terms of cutting force, surface roughness, and chipping thickness) are studied experimentally.

Introduction

Advanced ceramics (like alumina and silicon carbide) possess superior properties such as high strength, high hardness and rigidity at elevated temperatures, wear resistance, low thermal conductivity, and low chemical inertness. Due to these properties they find many applications in industry. Continuous research and development has resulted in new materials such as ceramic matrix composites (CMC) that combine reinforcing ceramic phases with a ceramic matrix. High alloy steels and refractory metals are being replaced by CMC materials because of their superior properties like high-temperature stability and high thermal-shock resistance [1-3]. Titanium alloys possess very high strength-to-weight ratio at elevated temperatures, exceptional corrosion resistance, and superior fatigue resistance, making them important materials in aerospace [4], automotive [5], chemical [6], medical and sporting goods [7] industries. However, these materials are difficult to machine. Excessive tool wear and high cost of machining are the major problems to their widespread applications. Rotary ultrasonic machining (RUM) is a non-conventional machining process that has potentials to drill holes in these hard-to-machine materials cost-effectively.

RUM is a hybrid machining process that combines the material removal mechanisms of diamond grinding and ultrasonic machining. Figure 1 is the schematic illustration of RUM. A rotary core drill with metal-bonded diamond abrasives is ultrasonically vibrated and fed toward the workpiece at a constant feedrate or a constant force (pressure). Coolant pumped through the core of the drill washes away the swarf, prevents jamming of the drill, and keeps it cool.

RUM was invented in 1964 [8]. The effects of control variables (rotational speed; vibration amplitude and frequency; diamond type, size and concentration; bond type; etc.) in RUM on its

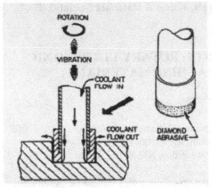
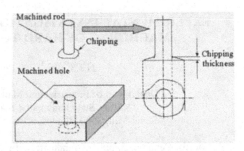

Figure 1. Illustration of RUM   Figure 2. Illustration of chipping thickness

performance (in terms of material removal rate, cutting force, surface roughness, etc.) have been investigated experimentally [9-15]. Efforts have also been made to develop models to predict the material removal rate in RUM from its control variables [16-18].

When drilling a hole with RUM, there is a tendency for the machined rod to break off before the machining finishes. This gives rise to edge chipping at the exit side of the hole. Figure 2 illustrates the edge chipping. The edge chipping degrades the quality of the machined hole. Research has been conducted to study the causes of edge chipping and find ways of reducing or eliminating it [1-3].

No other researchers have reported experimental investigations on RUM of alumina, ceramic matrix composite (CMC), silicon carbide, and titanium alloy (Ti-6Al-4V). This paper summarizes the experimental investigations on RUM of these four hard-to-machine materials, reporting the effects of spindle speed, feedrate, and ultrasonic vibration power on cutting force, surface roughness, and edge chipping.

Experimental Conditions and Procedures

Setup and Conditions

RUM tests were performed on an ultrasonic machine of Sonic Mill Series 10 (Sonic-mill, Albuquerque, NM, USA). Diamond core drills (N.B.R. Diamond Tool Corp., LaGrangeville, NY, USA) were used to drill holes in the workpieces. The drills with outer and inner diameter of 9.6 mm and 7.8 mm, respectively, consist of metal-bonded diamond grains of different mesh sizes. Mobilemet S122 water-soluble cutting oil (MSC Industrial Supply Co., Melville, NY, USA) was used as coolant (diluted by water with 1 to 20 ratio).

Table 1 shows the properties of the four workpiece materials. Table 2 shows the dimensions of the workpieces.

The process parameters investigated are spindle speed (rotational speed of the diamond core drill), feedrate (linear velocity of the drill in the direction normal to the workpiece surface), and ultrasonic vibration power (the percentage of the electrical power for ultrasonic vibration, controlling the amplitude of ultrasonic vibration).

140

Table 1 Properties of Workpiece Materials				
Property	Alumina	Silicon Carbide	CMC	Titanium Alloy
Tensile Strength (MPa)	129	3440	308	950000
Vicker's Hardness (VHN)	1190	2150	-	320
Density (Kg·m⁻²)	961	3100	740	4510
Melting Point (K)	56	2973	-	1941

Table 2 Workpiece Dimensions	
Workpiece Material	Workpiece Dimension (mm×mm×mm)
Alumina	25.4×25.4×6.3
Silicon Carbide	120×50×6
CMC	32×32×6.35
Titanium Alloy	115×85×7.2

Measurement of Output Variables

Three output variables were measured: cutting force, surface roughness, and chipping thickness. The cutting force along the feedrate direction was measured by a Kistler 9257 dynamometer (Kistler Instrument Corp, Amherst, NY, USA). The dynamometer was mounted atop the machine table and beneath the workpiece, as shown in Figure 3. The electrical signals from the dynamometer were transformed into numerical signals by an A/D converter. Then the numerical signals to measure the cutting force were displayed

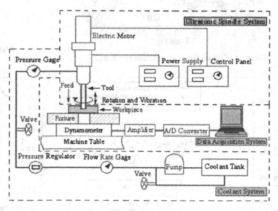

Figure 3. Experimental set-up

and saved on the computer with the help of National Instruments LabView (Version 5.1). The sampling frequency to obtain the cutting force signals was 100 Hz.

The surface roughness was measured on the machined hole surface with a surface profilometer (Mitutoyo Surftest-402, Mitutoyo Corporation, Kanagawa, Japan).

A digital video microscope of Olympus DVM-1 (Olympus America Inc., Melville, NY, USA) was utilized to inspect the edge chipping at the exit side of the machined hole. The hole quality was quantified by the thickness of the edge chipping formed on the machined rod, and measured with a vernier caliper (Mitutoyo IP-65, Mitutoyo Corporation, Kanagawa, Japan).

Experimental Results

Diamond Drilling vs. RUM

An experimental study has been conducted with and without ultrasonic vibrations (note that the drilling without ultrasonic vibrations is considered as diamond drilling) for CMC materials. Figure 4 shows the graph of cutting force vs. machining time. It can be seen that the fluctuations and maximum value of the cutting force for RUM are lower than those for diamond drilling. Figure 5 shows the cutting force data for two types of CMC materials and alumina. It can be

141

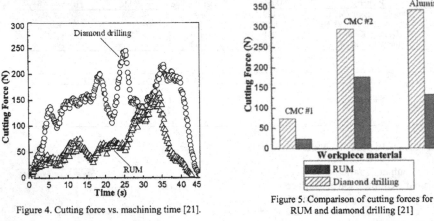

Figure 4. Cutting force vs. machining time [21].

Figure 5. Comparison of cutting forces for RUM and diamond drilling [21]

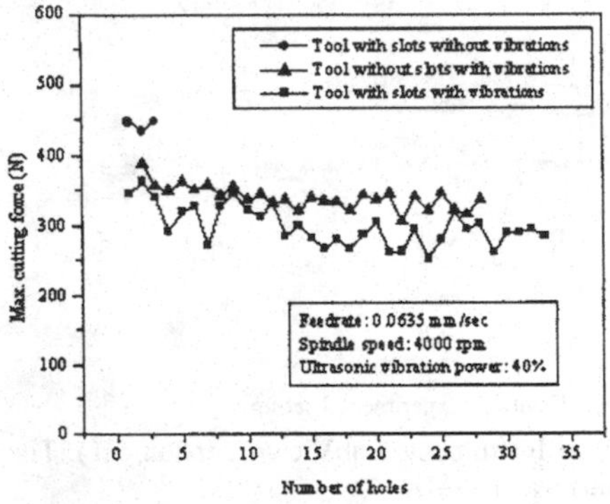

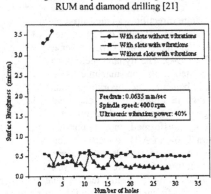

Figure 6. Cutting force comparison for three conditions

Figure 7. Surface roughness comparison for three conditions

observed that the maximum values of the cutting force for RUM are always lower than those for diamond drilling for all three materials.

Figure 6 shows the maximum cutting force vs. number of holes for titanium alloy. It can be observed that the maximum cutting force values for RUM are lower than those for diamond drilling.

Figure 7 shows the surface roughness vs. number of holes for titanium alloy. It can be seen that the surface roughness values for RUM are lower than those for diamond grinding.

Effects of Spindle Speed

Table 3 summarizes the effects of spindle speed. It can be seen that, as spindle speed increases, the cutting force decreases for alumina, silicon carbide, and titanium alloy, but increases for CMC; the surface roughness decreases for alumina, silicon carbide, and titanium alloy; and the chipping thickness decreases for alumina and CMC.

142

Table 3 Effects of Spindle Speed

Performance	Alumina	Silicon Carbide	CMC	Titanium Alloy
Cutting Force	Decrease	Decrease	Increase	Decrease
Surface Roughness	Decrease	Decrease	-	Decrease
Chipping Thickness	Decrease	-	Decrease	-

Table 4 Effects of Feedrate

Performance	Alumina	Silicon Carbide	CMC	Titanium Alloy
Cutting Force	Increase	Increase	Increase	Increase
Surface Roughness	Increase	Increase	-	Increase
Chipping Thickness	Increase	-	Increase	-

Table 5 Effects of Ultrasonic Vibration Power

Performance	Alumina	Silicon Carbide	CMC	Titanium Alloy
Cutting Force	Decrease	No Change	-	Decrease Then Increase
Surface Roughness	Increase	Decrease	-	Decrease
Chipping Thickness	-	-	Increase	-

Effects of Feedrate

Table 4 summarizes the effects of feedrate. It can be seen that, as feedrate increases, the cutting force increases for all four materials; the surface roughness increases for alumina, silicon carbide, and titanium alloy; and the chipping thickness increases for alumina and CMC.

Effects of Ultrasonic Vibration Power

Table 5 summarizes the effects of ultrasonic vibration power. It can be seen that the cutting force behaves differently for different materials. As the ultrasonic vibration power increases, the cutting force decreases for alumina, remains almost constant for silicon carbide, and decreases initially and then increases for titanium alloy; the surface roughness increases for alumina but decreases for silicon carbides and titanium alloy; and the chipping thickness increases for CMC.

Summary

1. The cutting force for rotary ultrasonic machining is lower than that for diamond drilling.

2. The increase in spindle speed, feedrate, and ultrasonic vibration power has different effects on cutting force, surface roughness, and chipping thickness for different materials.

Acknowledgements

The authors gratefully extend their acknowledgements to Mr. Bruno Renzi at N.B.R. Diamond tool Corp. for supplying the diamond core drills, Mr. James Jaskowiak at the Ferro Corporation for providing the alumina materials, Mr. Dean Owens at Saint-Gobain Ceramics for providing the silicon carbide materials, Mr. Ron Kestler at Aircraft Wheels & Brakes Goodrich Corp. for supplying the ceramic matrix composite panels, and Mrs. Tanni Sisco and Mr. Antonio Micale at Boeing Company for supplying the titanium alloy materials.

References

[1] D.W. Richerson, Ceramic matrix composites, Composite Engineering Handbook, Marcel Dekker Inc., New York, 1997.

143

[2] K. Okamura, Ceramic matrix composites (CMC), Journal of the Japan Society of Composite Materials 4 (3) (1995) 247-259.

[3] D.W. Freitag, D.W. Richerson, Ceramic matrix composites in opportunities for advanced ceramics to meet the needs of the industries of the future, Advanced Ceramic Association and Oak Ridge National Laboratory, Oak Ridge, Tennessee, 1998.

[4] R. Boyer, Overview on the use of titanium in the aerospace industry, Materials Science and Engineering A: Structural Materials: Properties, Microstructure and Processing 213 (2) (1996) 103-114.

[5] Y. Yamashita, I. Takayama, H. Fujii, T. Yamazaki, Applications and features of titanium for automotive industry, Nippon Steel Technical Report 85 (2002) 11-14.

[6] N. Orr, Industrial application of titanium in the metallurgical industries, and chemical light metals, Proceedings of Technical Sessions at the 111th American Institute of Mining, Metallurgical and Petroleum Engineers, Annual Meeting, AIME, Warrendale, PA (1982), pp. 1149-1156.

[7] F. Froes, Titanium sport and medical application focus, Materials Technology 17 (1) (2002) 4-7.

[8] P. Legge, Ultrasonic drilling of ceramics, Industrial Diamond Review 24 (278) (1964) 20-24.

[9] P. Legge, Machining without abrasive slurry, Ultrasonics (1966) 157-162.

[10] M. Kubota, Y. Tamura, N. Shimamura, Ultrasonic machining with a diamond impregnated tool, Bulletin of Japan Society of Precision Engineering 11 (3) (1977) 127- 132.

[11] A. Markov, I. Ustinov, A study of the ultrasonic diamond drilling of non-metallic materials, Industrial Diamond Review (1972) 97-99.

[12] A. Markov, Ultrasonic drilling and milling of hard non-metallic materials with diamond tools, Machine and Tooling 48 (9) (1977) 45-47.

[13] P. Petrukha, Ultrasonic diamond drilling of deep holes in brittle materials, Russian Engineering Journal 50 (10) (1970) 70-74.

[14] D. Prabhakar, Machining advanced ceramic materials using rotary ultrasonic machining process, M.S. Thesis 1992, University of Illinois at Urbana -Champaign.

[15] D. Prabhakar, P.M. Ferreira, M. Haselkorn, An experimental investigation of material removal rates in rotary ultrasonic machining, Transactions of the North American Manufacturing Research of SME 10 (1992) 211-218.

[16] D. Prabhakar, Z.J. Pei, P.M. Ferreira, M. Haselkorn, A theoretical model for predicting material removal rates in rotary ultrasonic machining of ceramics, Transactions of the North American Manufacturing Research Institution of SME 21 (1993) 167–172.

[17] Z.J. Pei, P.M. Ferreira, Modeling of ductile mode material removal in rotary ultrasonic machining, International Journal of Machine Tools and Manufacture 38 (10–11 (1998) 1399–1418.

[18] Z.J. Pei, D. Prabhakar, P.M. Ferreira, M. Haselkorn, A mechanistic approach to the prediction of material removal rates in rotary ultrasonic machining, Journal of Engineering for Industry (117 (2) (1995) 142–151.

[19] Y. Jiao, W.J. Liu, Z.J. Pei, X.J. Xin, C. Treadwell, Study on edge chipping in rotary ultrasonic machining on ceramics: an integration of designed experiment and FEM analysis, Journal of Manufacturing Science and Engineering 127 (4) (2005) 752–758.

[20] Y. Jiao, P. Hu, Z.J. Pei, C. Treadwell, Rotary ultrasonic machining of ceramics: design of experiments, International Journal of Manufacturing Technology and Management 7 (2-4) (2005) 192-206.

[21] Z.C. Li, Y. Jiao, T.W. Deines, Z.J. Pei, C. Treadwell, Rotary ultrasonic machining of ceramic matrix composites: feasibility study and designed experiments, International Journal of Machine Tools and Manufacture 45 (12–13) (2005) 1402–1411.

Materials Processing under the Influence of External Fields
Edited by Qingyou Han, Gerard Ludtka, Qijie Zhai
TMS (The Minerals, Metals & Materials Society), 2007

NON-CONTACT ULTRASONIC TREATMENT OF METALS IN A MAGNETIC FIELD

John B. Wilgen[1], Roger A. Kisner[1], Roger A. Jaramillo[2], Gerard M. Ludtka[2],
Gail Mackiewicz-Ludtka[2]

[1]Engineering Science and Technology
[2]Materials Science and Technology
Oak Ridge National Laboratory
1 Bethel Valley Rd, Oak Ridge, Tennessee, 37831, USA

Keywords: ultrasonic, EMAT, magnetic, field, non-contact, aluminum, A356

Abstract

A high-field EMAT (Electromagnetic Acoustical Transducer) has been used for non-contact ultrasonic processing of aluminum samples during solidification. The magnetic field for the EMAT is supplied by a high-field (20 Tesla) resistive magnet, and the current is provided by an induction coil. This resulted in a highly efficient EMAT that delivered 0.5 MPa of acoustic drive to the surface of the sample while coupling less than 100 watts of incidental induction heating. The exceptionally high energy efficiency of the electromagnetic transducer is due to the use of high magnetic field, which reduces the current needed to achieve the same acoustic pressure. In the initial experiments, aluminum samples of A356 alloy were heated to the liquid state and allowed to solidify at a controlled cooling rate while subjected to the non-contact ultrasonic stimulation (0.5 MPa @ 165 kHz) provided by an induction coil located within the bore of a 20-T resistive magnet.

Introduction

Ultrasonic processing of materials in both the melt and solid phase is proving to be highly beneficial to material properties of metallic alloys [1]. In the melt phase, acoustic treatment can be used to enhance diffusion, dispersion, and dissolution processes, resulting in improvements in the cleaning, refining, degassing, and solidification of the melt. Ultrasonic processing can be used to assist in grain refinement and to minimize segregation during solidification. Degassing with ultrasonics has resulted in reduced gas concentration, higher density, and improved mechanical properties. It has been demonstrated that non-dendritic structures can be produced with ultrasonic cavitation treatment, resulting in increased plasticity and enhanced strength. In the solid state phase, ultrasonic treatment could potentially be utilized to minimize residual stress, accelerate phase transformation processes, enhance nucleation and growth during phase transformations, enhance diffusive processes by enhancing the mobility of diffusing species, and enhance processes that have a threshold activation energy. The method can be coupled with high-field magnetic processing, [2,3] but can also be used where only ultrasonic treatment is beneficial.

Commercially available ultrasonic processing systems require direct contact with the melt, resulting in undesirable chemical interactions when the acoustic probe/horn is inserted directly into the molten material. Ultrasonic transducers are limited in temperature range, and therefore must be thermally isolated from high-temperature environments by using an acoustical waveguide, or horn. Acoustic impedance mismatches between the transducer and the waveguide, as well as between the waveguide and the melt can limit the transfer of energy. Various types of probe coatings have been investigated in an effort to minimize the chemical interactions of the probe surface with the melt. In addition, the localized nature of the horn probe results in a very non-uniform distribution of acoustical energy within the melt crucible. The ability to couple acoustic energy efficiently via a non-contacting method would overcome a huge technological barrier to the more widespread use of ultrasonic processing.

Theory

When induction heating is applied in a high magnetic field environment, the induction heating coil is typically configured in such a way that high intensity ultrasonic treatment occurs naturally. The resulting configuration is that of a highly effective electromagnetic acoustical transducer (EMAT). This approach is unique because the magnetic field strength of the high-field magnet can greatly exceed that used in a typical EMAT device. As a consequence, the synergistic interaction of a high surface current density (induced via induction heating) with the steady-state high magnet field results in an especially effective method for creating a high energy density acoustic environment. The exceptionally high energy efficiency of the resulting electromagnetic transducer is due to the use of high magnetic field, which greatly reduces the current needed to achieve the same acoustic pressure. This provides an efficient non-contact method for applying high-intensity ultrasonic energy to the processing of metals. Furthermore, the applied ultrasonic excitation can be uniformly distributed over most of the surface of the metal sample.

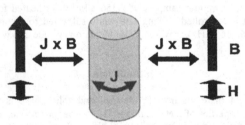

Figure 1. Induction heating in a high-field magnet, showing the applied H-field of the induction coil, the induced surface current in the sample (J), the static field of the magnet (B), and the resulting JxB force. The H-field of the induction heating coil is insignificant (μH << B) by comparison with the static B-field of the high-field magnet.

If the axis of a helical induction heating coil is aligned with the static magnetic field of a high-field magnet, the azimuthally-directed surface current induced in the process metal interacts with the static field of the magnet. The result is a large oscillatory electromagnetic force, or pressure, that acts directly on the metal surface, at the induction heating frequency. In cylindrical coordinates, the force is in the radial direction. Figure 1 shows the applied H-field of the induction coil, the induced azimuthally-directed surface current (J), the static externally applied magnetic field (B), and the resulting radially-directed electromagnetic force (J x B). The acoustic driving force is bi-directional, alternately compressing and stretching (tensioning) the sample in the radial direction. In liquids, the later leads to cavitation that can be very beneficial for ultrasonic processing of the melt phase. The acoustic pressure can be quite substantial since both the induced surface current and the static magnetic field are quite large.

The effectiveness of acoustical excitation can be enhanced if the frequency happens to coincide with a natural resonant frequency of the sample. Because of the large mismatch in the acoustic impedance at a material-air interface, most of the acoustic energy can be trapped within the sample and the sample container, forming an acoustical resonator. If the acoustic drive frequency is chosen to match a natural resonant frequency of the sample, then the peak acoustic pressure in the resonator can be enhanced by a factor that is equal to the quality factor of the resonator. Quality factors for liquid metal columns with large length-to-diameter ratios are expected to be in the range of 10-100 [4]. Although a somewhat smaller quality factor might be anticipated for a typical experimental configuration, due consideration should be given to the potentially advantageous use of acoustical resonances.

Using this high-field EMAT method, non-contacting ultrasonic treatment can be applied to the processing of metal alloys in either the solid and melt phase. Molten metals can be contained in a non-metallic ceramic crucible that is readily penetrated by the electromagnetic induction fields. Ultrasonic treatment in the solid phase can be achieved either at near-ambient temperatures, or at elevated temperature. When applied under high temperature conditions, temperature control can be achieved via the induction heating process, whereas at ambient temperatures, gas cooling might be required to remove the incidental heat coupled by the induction system.

Experimental

Aluminum samples processed ultrasonically with the high magnetic field EMAT were compared with samples processed in the absence of EMAT excitation (no-field), but with the same thermal treatment. Aluminum samples of A356 alloy were inductively heated and melted in an alumina crucible located within the 8-inch diameter bore of the high-field solenoidal resistive magnet, in an argon atmosphere. The sample temperature was measured using a sheathed thermocouple inserted into a thin-walled stainless steel well that allowed the thermocouple to be withdrawn after solidification of the melt. When the melt temperature reached 700 C, the magnetic field was ramped up to the desired value, usually either 9 or 18 Tesla. Since the ultrasonic pressure resulting from the EMAT effect is proportional to the externally applied B-field, the ultrasonic excitation of the sample increased linearly during the ramp-up of the magnetic field. The samples were then allowed to solidify at a controlled cooling rate while being subjected to the non-contact ultrasonic stimulation (0.5 MPa @ 165 kHz) provide by the EMAT. Forced flow helium gas was used to cool down and solidify the samples while continuing to power the induction coil at the maximum current level. This procedure maintains a constant level of ultrasonic drive throughout the cool down and solidification of the sample, as both the induction current and the external field were held constant during the cool down phase. The cooling rate was controlled by adjustment of the flow rate of the helium coolant gas.

In Figure 2, two sets of cooling curves are shown, corresponding to two different flow rates of coolant gas. Each set includes the measured sample temperature data for two different values of magnetic field including a no-field case and an 18 T field case. Each plot shows two inflections in the cooling rate due to exothermic events. These recalescence events, one at 610 C and the other at 570 C, are associated with the onset and completion of alloy solidification, respectively. These temperatures are consistent with liquidus and solidus temperatures of 615 C and 575 C predicted by Thermocalc, Al-DATA database [5], for this alloy. For the slower cooling rate achieved with a 10 psi helium feed pressure, solidification is completed in approximately 11 minutes. To effect a faster cooling rate, the helium feed pressure was increased to 40 psi, which

147

decreased the cooling and solidification time by more than a factor of two, to approximately 4 minutes. For the 18 Tesla case, the ultrasonic drive pressure is the maximum achievable value of approximately 0.5 MPa. For the no-field case, there remains a residual EMAT effect arising from the induction heating along, but this is much weaker, and at twice the induction frequency.

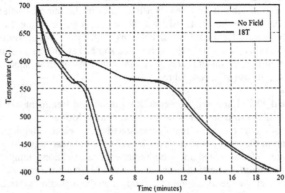

Figure 2. Cooling curves for two different flow rates of helium coolant gas

Results and Discussion

Aluminum samples processed with the high magnetic field EMAT were compared with samples process in the absence of magnetic field (no EMAT) but with the same thermal treatment. Based on visual appearance, samples processed with EMAT stimulation showed notably improved surface conditions as compared with samples solidified in the absence of EMAT stimulation under no-field conditions. This difference is illustrated by the photos shown in Figure 3. The processed samples were approximately 35 mm in diameter, and about 52 mm in length.

Figure 3. The sample on the left (a) was processed in the absence of EMAT stimulation under no-field conditions, whereas the sample on the right (b) was solidified with maximum EMAT stimulation using an external magnetic field of 18 T.

Optical micrographs reveal that there were obvious differences in microstructure uniformity for the samples that were subjected to ultrasonic processing using the high-field EMAT.. Micrographs for a no-field sample, shown in Figure 4, reveal a very obvious difference in microstructure between the top and the bottom of the sample, suggesting a segregation of the Silicon, which is manifest in different amounts of primary alpha (~pure Al) versus eutectic (alpha plus Si-rich phase). By comparison, similar micrographs for a high-field EMAT sample, shown in Figure 5, do not show any variations in microstructure uniformity between the top and

148

bottom of the sample. This sample was processed with maximum EMAT stimulation using an external magnetic field of 18 T. Additional analysis of the samples is needed to provide a more complete assessment of the differences attributable to high field EMAT processing.

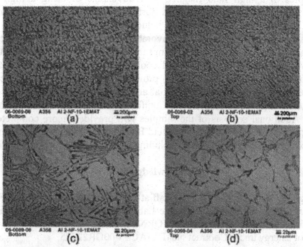

Figure 4. Optical micrographs of no-field sample comparing low resolution images near the bottom (a) and top (b) of the sample, as well as 10x higher resolution images near the bottom (c) and top (d) of the sample.

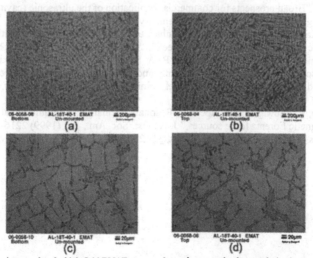

Figure 5. Optical micrographs of a high-field EMAT processed sample comparing low resolution images near the bottom (a) and top (b) of the sample, as well as 10x higher resolution images near the bottom (c) and top (d) of the sample.

Summary and Conclusions

A high-field EMAT has been used to effect non-contact ultrasonic solidification processing of aluminum samples in a high field magnet. An advantage of using a high magnetic field environment is that this allows the induced surface current in the sample to be decreased proportionately. As a result, the incidental induction heating associated with the use of the EMAT transducer is greatly reduced, which favorably affects the energy efficiency of the EMAT approach. In the initial experiments, aluminum samples of A356 alloy were heated to the liquid state and allowed to solidify at a controlled cooling rate while subjected to the non-contact ultrasonic stimulation (0.5 MPa @ 165 kHz) provided by an induction coil located within the bore of a 20-Tesla resistive magnet. An initial analysis of the samples produced during these experiments shows improved microstructure uniformity and a reduction in porosity near the radial surface. Additional analysis of the samples will provide a more complete assessment of the benefits attributable to non-contact high-field EMAT processing, and additional experiments are needed to explore the parameter space, including optimization of the ultrasonic frequency.

Acknowledgement

We acknowledge Dr. Bruce Brandt and the staff at the National High Magnetic Field Laboratory for their support. Research sponsored by the Laboratory Directed Research and Development Program of Oak Ridge National Laboratory (ORNL), managed by UT-Battelle, LLC for the U. S. Department of Energy under Contract No. DE-AC05-00OR22725.

References

1. G. I. Eskin, "Broad prospects for commercial application of the ultrasonic (cavitation) melt treatment of light alloys," Ultrasonic Sonochemistry, 8 (2001) 319-325.
2. G.M. Ludtka, et. al.,"In-situ Evidence of Enhanced Phase Transformation Kinetics Due to a High Magnetic Field in a Medium Carbon Steel,", Scripta Materialia, 51 (2004) 171-174.
3. R. A. Jaramillo, S. S. Babu, G. M. Ludtka, R. A. Kisner, J. B. Wilgen, G. Mackiewicz-Ludtka, D. M. Nicholson, S. M. Kelly, M. Murugananth, and H. K. D. H. Bhadeshia, "Effect of 30 Tesla Magnetic Field on Transformations in a Novel Bainitic Steel", Scripta Materialia, 52 (2005) 461-466.
4. S. Makarov, R. Ludwig, and D. Apelian, "Resonant oscillation of a liquid metal column driven by electromagnetic Lorentz force sources," J. Acoust. Soc. Am., 105 (1999) 2216-24.
5. N. Saunders, J. Japanese Inst. Light Metals, 51 (2001) 141.

Materials Processing under the Influence of External Fields
Edited by Qingyou Han, Gerard Ludtka, Qijie Zhai
TMS (The Minerals, Metals & Materials Society), 2007

NUMERICAL SIMULATION FOR MOLD-FILLING AND COUPLED HEAT TRANSFER PROCESSES OF TITANIUM SHAPED CASTINGS UNDER CENTRIFUGAL FORCES

Daming Xu, Hongliang Ma, Jingjie Guo
School of Materials Science and Engineering, Harbin Institute of Technology
E-mail address: damingxu@hit.edu.cn

Keywords: Centrifugal casting, Mold-filling, Heat transfer, Numerical simulation

Abstract

Centrifugal casting technique is usually used to produce high quality castings of special metals, e.g. titanium alloys, with thin-sections and complicated shapes. Due to the more complicated centrifugal-mold-filling behaviors compared with that of gravity casting, computer modeling methods are needed to reveal the mold-filling behaviors under a centrifugal force field, and to obtain a properly technological control for a centrifugal casting process. A mathematical model and a computer software system, named 3DLX-WSIM, for a general 3D-simulation purpose for centrifugal field casting processes have been developed by the authors. This article exhibits the recent computer simulation results for the mold-filling and the coupled heat-transfer behaviors on a more practical-oriented titanium centrifugal-casting system.

Introduction

In order to efficiently enhance the mold-filling abilities for titanium alloy melts, a special casting technique, which is under centrifugal forces and could be termed as centrifugal casting, is usually adopted to obtain a high quality casting with thin-sections and complicated shapes [1]. However, due to the dramatic variations in the centrifugal forces with rotating rates and the distances to the rotating axis, the alloy melts may exhibit much more complicated mold-filling behaviors compared with that only under the gravity [2]. On the other hand, such mold-filling and the coupled heat-transfer/solidification behaviors are difficult to be experimentally determined, due to the rotations and the opacities of both the metallic melts and the mold materials. Therefore, it is useful to develop a computer simulation technique to exhibit the mold-filling behaviors under a centrifugal force field, and to obtain a rationally technological control for a centrifugal casting process.

A mathematical model and a computer codes system for a general 3D-simulation purpose for both centrifugal-field and gravity casting processes have been developed and recently modified by the authors (e.g. Coriolis force has been added to the software), which is renamed as 3DLX-WSIM for the new computer-modeling version. Aims of this article is to investigate the influences both of a centrifugal force and the Coriolis force on the mold-filling and the coupled heat-transfer behaviors in a more practical-oriented titanium centrifugal-casting system, as well as to present the availabilities of the authors' new-version computer codes.

Computation Model

A centrifugal casting configuration is illustrated in Figure 1(a), in which the mold is assumed to rotate around the vertical axis z at a constant rotating rate, ω_0. When a gravity-driven downwards melt inlet splashing onto the bottom rotating mold wall, the filling melt is gradually accelerated to gain a rotating rate, ω_n, with further contacting with the rotating mold walls of the casting cavity. Such rotating melt in turn posses a centrifugal force, as a result, should have a much stronger mold-filling ability than just under the gravity.

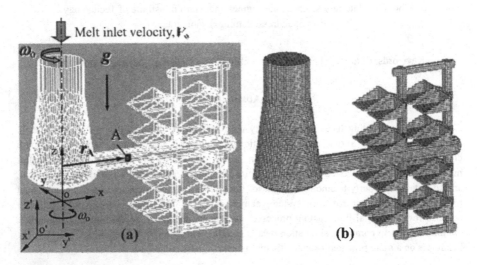

Figure 1 A centrifugal casting configuration (a); and the mesh pattern for the casting cavity (total casting mesh number is 993,836) (b).

In the present modeling, the governing equations for the centrifugal mold-filling and the coupled heat-transfer processes of an alloy melt are formulated in the constantly rotating coordinate system, x-y-z, as shown in Figure 1(a). The model may take the following equation form.

Momentum Transfer Equation

$$\frac{\partial v}{\partial t} + \nabla(v \cdot v) = \gamma \nabla^2 v - \frac{1}{\rho_L} \nabla P + g + \omega_n^2 r_A - 2\omega_0 \times v \tag{1}$$

Continuity Equation

$$\nabla \cdot v = 0 \tag{2}$$

Volume of Fluid (VOF) Equation

$$\frac{\partial F}{\partial t} + \nabla(vF) = 0 \tag{3}$$

152

Heat Energy Transfer Equation

$$\frac{\partial T}{\partial t} + \nabla\left(vT\right) = \alpha\nabla^2 T + \frac{h}{c_p}\frac{\partial f_s}{\partial t} \tag{4}$$

In the above model equations, the vector symbol $v = v_x i + v_y j + v_z k$ represents the relative melt-flow velocity in the rotating coordinate, x-y-z, system. In the momentum transfer Equation (1), the fourth term on the right-hand-side, i.e. $a_n = \omega_n^2 r_A$, represents the centrifugal acceleration, where r_A is the vector distance from the rotating axis pointing to the center of a concerned control melt-volume A, see Figure 1(a); The absolute angular velocity of the control volume A, ω_n, i.e. in the laboratory (static) coordinate x'-y'-z', can be expressed as [3, 4]:

$$\omega_n = \omega_o + (v_y \cos\theta - v_x \sin\theta)/|r_A| \tag{5}$$

where, θ is the angle between the x-axis and the projection of vector r_A on the x-o-y plane. The last term on the right-hand-side of Equation (1) represents the Coriolis acceleration, resulted from a kind of inertia force.

SOLA-VOF algorithm [5, 6] has been widely used in various casting-CAE computer codes for casting mold-filling computations. In the present work, this numerical solution technique is also adopted to calculate the centrifugal mold-filling procedures (VOF-distribution) and the relative velocity field etc, but in a rotating coordinate system, based on the heat-transfer-coupled mathematical model of Equations (1)-(5) and a centrifugal casting configuration as shown by Figure 1(a).

All the numerical computations are performed with an authors' developed software for a general 3D-simulation purpose for various casting processes both only under the gravity (setting $\omega_o \equiv 0$.) and in a rotating-casting mold system, as illustrated in Figure 1(a). The program system consists of three major modules of pre-processing, numerical computation, and post-processing, which is more recently improved and renamed as 3DLX-WSIM for the new system version [7]. The pre-/post-processing modules provide basic functions of 3D solid-reconstruction and meshing for a practical precision casting system, and 3D color-image presentations for the numerical simulation results of different centrifugal mold-filling and coupled heat-transfer processes. All the calculations by the computing module and the color-image presentations by the pre-/post-processing modules can be performed on a PC-computer. Figure 1(b) shows a meshing image output by the authors' computer codes for a practical-oriented centrifugal titanium-casting cavity, corresponding to the centrifugal casting configuration shown by Figure 1(a). This mesh model is constructed by nearly one million meshed elements, and will be used for the sample computations in the following section.

The centrifugal casting system to be simulated, as shown by Figure 1, has a complex gating system of a rotating inlet-melt container (a special sprue), a horizontal runner and a vertical gating frame. 16 castings are attached to the vertical gating frame via a short inner-gate in four columns and four layers. The maximum rotating radius of the casting system for the filling melt is about 650 mm. The model casting alloy is taken as Ti-6Al-4V, the used properties data of which for the present computer simulations is listed in Table I. The pouring temperature and the downwards inlet-velocity of the alloy melts are assumed to be 1700°C and 1000mm/s, respectively. During a centrifugal casting process, the casting mold is assumed to turn around the z-axis, which superposes with that of the downwards melt-inlet, at a constant angular velocity, ω_b. Effects of both the different magnitudes and directions of the mold-rotating rate ω_b on the centrifugal mold-filling behaviors will be tested in the present computer modeling.

Table I   The Properties Data Used in the Present Computer Simulations for Ti-6Al-4V Alloy

Kinetic viscosity, γ [mm^2/s]	Density, ρ_L [g/mm^3]	Specific heat, c_P [J/g·°C]	Heat conductivity, λ [W/mm·°C]	Latent heat, h [J/g]	Liquidus temperature, [°C]	Solidus temperature, [°C]
0.954	4.44×10^{-3}	0.628	0.0155	375.72	1635	1522

Computational Results and Discussions

In a previous modeling for centrifugal mold-filling, the authors' emphases were placed on the impacts of the centrifugal forces, i.e. the $\omega_n^2 r_A$ term in Equation (1), while the Coriolis force effects were neglected [3, 4, 7-9]. Despite such neglecting, a full model simulation for a similar centrifugal casting system, carried out in Ref. [10], shows that the calculated results are basically similar. This indicates that the Coriolis force may exert less important influences on the mold-filling behaviors. More recently, the computing module of the 3DLX-WSIM system has been correspondingly revised for the full modeling by Equations (1)-(5). One of the major purposes of this work, therefore, is to investigate the actions of the Coriolis term in a more practical centrifugal casting process, by comparing the calculated results obtained from the previous and the newly revised computer codes, respectively.

Centrifugal Force Impacts—Previous Modeling Version

Figure 2 shows a comparison between the simulation results for the centrifugal mold-filling and the coupled heat transfer behaviors at ω_o= 90rpm and that only under the gravity (ω_o= 0rpm), obtained from the previous computing codes same as that used in Refs. [3, 4, 7-9]. In all the 3D image-presentations of the present paper, output from the 3DLX-WSIM system, the different colors in the filling melt domains represent the corresponding temperatures of the alloy melts, as defined by the color-temperature (C-T) scale shown in Figure 2(a).

Again, the computational results comparison of Figure 2 indicates that, for the more complicated centrifugal casting system, the alloy melt with a centrifugal force field may possess of much stronger mold-filling ability than that only under the gravity. It is also interesting to see that under a centrifugal force field, the alloy melt tends to fill-up the farthest vertical gate first, then to fill the casting cavities progressively in a backward direction (roughly towards to the rotating axis). Such mold-filling processes are principally driven by the corresponding paraboloidal pressure profiles in the alloy melt, built up by the centrifugal acceleration, $a_n = \omega_n^2 r_A$ [2]. These mold-filling behaviors are similar to the previously simulated results [3, 8].

154

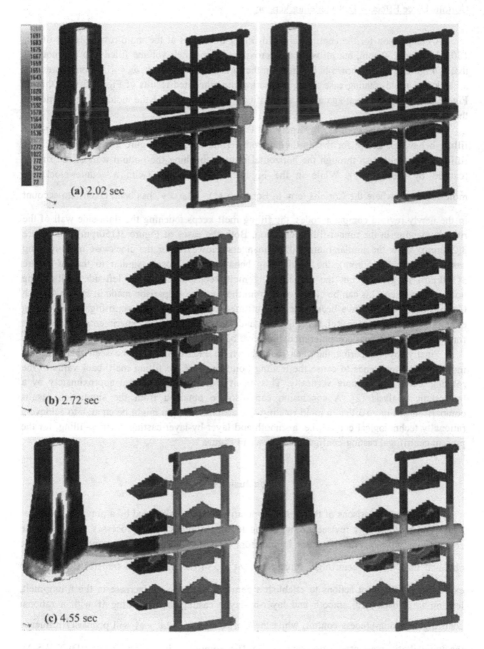

Figure 2   Comparisons of the simulation results for the centrifugal mold-filling and the coupled heat transfer processes (ω_o=90rpm) at different times (Left column) with that only under the gravity (ω_o=0rpm) (Right column).

A comparison for the centrifugal mold-filling behaviors at the mold-rotating rates of 90, 150, -150 and 250 rpm, but all with an approximate same melt-volume filled into the mold as that of Figure 2(b), are shown in Figure 3. In the figure, the minus rate $\omega_b = -150$ rpm represents a clockwise mold-rotating case. First, comparing the simulated results of Figure 3 with that of Figure 2(b), it can be seen again that a centrifugal force do enhance the mold-filling abilities of the alloy melt.

Furthermore, the Figure 2(b/90rpm) and Figure 3(90rpm) exhibit almost same casting-part filling behaviors, except for that in the horizontal runners. In the Figure 2(b-90rpm) case, the filling alloy melts rush through the horizontal runner, touching the bottom wall of the runner (caused by the gravity). While in the Figure 3(90rpm) case (with a counter-clockwise mold-rotating), where the Coriolis term in Equation (1), $-2\omega_o \times v$, has been taken into account in the newly revised computer codes, the filling melt keeps touching the right-side wall of the runner, viewing in the runner-filling direction. Both the cases of Figure 3(150rpm) and Figure 3(250rpm) show the similar runner-filling manners. However, for the clockwise mold-rotating case of Figure 3(-150rpm), the mold-filling behaviors are closely similar to that of Figure 3(150rpm) in general, but the runner-filling melt keeps touching the left-side wall of the horizontal runner. This can be clearly seen from the further comparison made in Figure 4. Such Coriolis force effects have been directly confirmed by the Ti-6Al-4V centrifugal mold-filling experiments carried out in Ref. [1] (Figures 2 and 3 of Ref.[1]). More experimental verifications for the authors' previous computer modeling have been shown in Refs. [3, 4 and 9].

Finally, by comparing the each case shown in Figure 3, it can be seen that a higher mold-rotating rate tends to cause the leading boundaries of the filling melt, backwards to the rotating z-axis, to be more vertically. This result can also be predicted approximately by a hydrostatic analysis [2]. A conclusion can also be obtained from the simulation results comparison of Figure 3 that, a mold-rotating rate around 200 rpm might be enough to achieve a rationally technological control, i.e. a smooth and layer-by-layer casting-cavities-filling, for the present centrifugal casting-configuration shown in Figure 1.

Conclusions

From the comparisons of the present simulation results, calculated by a previous computer codes and by the newly revised full-modeling for centrifugal casting processes, as well as the comparisons with experiments made in and from Refs. [3, 4, 9 and 1], the conclusions can be obtained that: 1) The centrifugal force term, $\omega_n^2 r_A$, in the momentum transfer Equation (1) exerts the predominant actions to establish a paraboloidal pressure increase in the filling-melt, leading to a backward, smooth and layer-by-layer casting-cavities-filling if with a rational centrifugal-casting-process control, while the Coriolis term, $-2\omega_o \times v$, will primarily influence the initial horizontal-gate-filling manners; 2) The computer simulation system 3DLX-WSIM, developed by the authors, has been shown to be a useful and reliable tool to simulate a 3D centrifugal casting-process.

Figure 3 A surface view on the centrifugal mold-filling behaviors with different mold-rotating velocities and directions (minus ω_b rate standing for a clockwise mold-rotating).

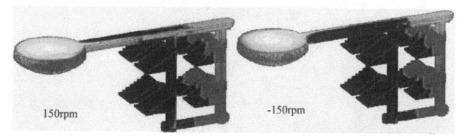

Figure 4 Comparison between the mold-filling behaviors in the horizontal runners with different mold-rotating directions at $|\omega_b| = 150$ rpm, showing the Coriolis force's effects (minus ω_b rate representing a clockwise mold-rotating).

References

1. K. Suzuki, K. Nishikawa, and S. Watakabe, "Mold Filling and Solidification during Centrifugal Precision Casting of Ti-6Al-4V Alloy", *Materials Transactions, JIM*, 37 (12) (1996), 1793-1801.
2. Y.A. Filin and A.S. Isaev, *Casting of New Alloys for Marine Vehicles* (Former Soviet Union: Л., Судостроение, 1971), 196-204.
3. Li Xin, "A Study on Numerical Simulation of Mold-filling and Solidification Process of Centrifugal Castings" (Master thesis, Harbin Institute of Technology, 2001), 15-41.
4. Xu, Daming et al., "Numerical Simulation of Mold-filling Behaviors under Centrifugal Forces", *Chinese J. of Mechanical Eng.*, 39 (3) (2003), 146-150.
5. C.W. Hirt and B.D. Nichols, "Volume of Fluid (VOF) Method for the Dynamics of Free Boundaries", *J. of Computational Phys.*, 39 (1981), 201-225.
6. B.D. Nichols, C.W. Hirt and R.S. Hotchkiss, "SOLA-VOF: A Solution Algorithm for Transient Fluid Flow with Multiple Free Boundaries", (Tech. Report LA-8355, Los Alamos Scientific Lab, 1980).
7. Ma Hongliang, "Improvements on the Software System for Simulations of Centrifugal Mold-Filling and Heat-Transfer Processes of Titanium Castings" (Master thesis, Harbin Institute of Technology, 2006).
8. Daming Xu et al., "Computer Simulation for Centrifugal Mold Filling of Precision Titanium Castings", *China Foundry*, 1 (1) (2004), 53-57.
9. Xu, Daming et al., "Mold Filling Behaviors of Melts with Different Viscosity under Centrifugal Force Field", *J. of Mater. Science and Technology*, 18 (2) (2002), 149-151.
10. Zeng Xingwang, "Simulation to Practical Flowing Process of Centrifugal Casting" (Master thesis, Huazhong University of Science &Technology, 2004), 50-53.

Materials Processing
Under the
Influence of External Fields

Session IV

Materials Processing under the Influence of External Fields
Edited by Qingyou Han, Gerard Ludtka, Qijie Zhai
TMS (The Minerals, Metals & Materials Society), 2007

INTEGRATED MULTIPHYSICS AND MULTISCALE MODELING OF ELECTROMAGNETICALLY-ASSISTED MATERIALS PROCESSING SYSTEMS

Ben Q. Li

Department of Mechanical Engineering
University of Michigan
Dearborn, MI 48128

Keywords: electromagnetics, multiphysics, multiscales, solidification

Abstract

This paper discusses the development of integrated multiphysics and multiscale numerical models for electromagnetically-assisted materials processing systems. Mathematical formulations and computational methodologies used to develop these integrated models are presented. While a fully integrated coupled solution across the atomistic scale through continuum is theoretically possible, a more feasible approach for developing models for practical applications would involve a mixture of sequential and mutual coupling across one or two scales. The use of an integrated multiscale and multiphysics model is illustrated through an example taken from solidification processing of melts with an externally applied induction stirring, where the evolution of solidification microstructure formation is predicted simultaneously along with the transient development of the electromagnetic, fluid flow and thermal fields.

Introduction

Electromagnetically-assisted materials processing systems involve complex interacting physical phenomena, which may take place at spatial and temporal scales in a great disparity. A notable example of this type is the solidification of melts with electromagnetic stirring. In such a system, atoms in the liquid phase below the meting point are incorporated into the solid phase by passing through an energy barrier. Different crystal structures and/or grain orientations have different barriers, which affect this atomistic transition from the liquid to the solid phase. The growth of the crystals is a result of the continuous deposition of these atoms onto the solid phase. These microscopic events are further influenced by the temperature field which is a continuum phenomenon, or more precisely a statistically averaged microscopic phenomenon. Figure 1 illustrates these interacting multiscale phenomena in a solidifying pool. Such complex multiscale solidification phenomena are further complicated by an application of an external electromagnetic field, which is intended to produce a desired solidification microstructure through modifying the temperature and fluid flow fields.

Understanding of these multiphysics and multiscale phenomena is essential for the control and optimization of existing eletromagnetically-assisted solidification processes as well as for the design of new processes for mission-critical applications. Prediction of these interactive multiscale phenomena presents a challenging task. The rapid advance in computing power and

161

software algorithms in recent years, however, has made it possible to start looking into the integrated solution of these types of problems that involve multiphysics and multiscales.

This paper discusses some recent work on the subject of integrated modeling of multiphysics and multiscale phenomena in materials processing systems in the presence of external fields. The discussion is presented in the context of during solidification processing with electromagnetic stirring. The modeling approach, of course, can be adapted to model a solidification process with other types of external fields. In what follows, we present the basic mathematical description of the complex multiscale physical events, followed by the computational methodologies for modeling these events and by the strategies by which these methodologies are coupled to develop an integrated comprehensive model.

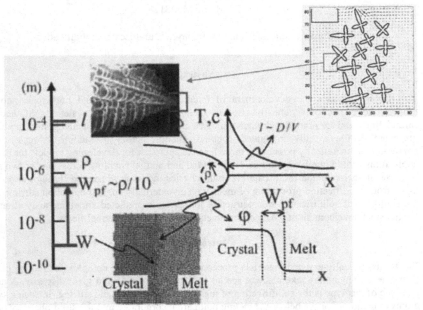

Figure 1. Multiscale phenomena in liquid-solid transition [1].

It is worth noting that some well studied materials processing systems such as electromagnetic levitation, electrostatic levitation, induction stirring, microwave sintering and magnetic damping may also fall into this category of problems, where interactive multiphysics plays an essential role and yet localized phenomena may or may be as important, depending upon specific applications. The multiphysics involved in these systems may also be modeled using the integrated computational methodology presented below, wherever applicable.

Macrosopic Description of Multiphysics

In the context of solidification under active electromagnetic stirring, the mathematical statement of the macroscopic phenomena consists of the Maxwell equations and the momentum and thermal balance equations [2-5],

$$\nabla \times \mathbf{E} = -\frac{\partial \mathbf{B}}{\partial t} \tag{1}$$

$$\nabla \times \mathbf{H} = \mathbf{J} + \frac{\partial \mathbf{D}}{\partial t} \tag{2}$$

$$\nabla \cdot \mathbf{B} = 0 \tag{3}$$

$$\nabla \cdot \mathbf{D} = 0 \tag{4}$$

$$\mathbf{J} = \sigma (\mathbf{E} + \mathbf{u} \times \mathbf{B}) \tag{5}$$

$$\rho \frac{\partial \mathbf{u}}{\partial t} + \rho \mathbf{u} \cdot \nabla \mathbf{u} = -\nabla p + \nabla \cdot \left(\mu_{eff} \nabla \mathbf{u} \right) - \rho \beta \mathbf{g}(T - T_r) + \mathbf{J} \times \mathbf{B} + \mathbf{F}_{ext} \tag{6}$$

$$\nabla \cdot \mathbf{u} = 0 \tag{7}$$

$$\rho C \frac{\partial T}{\partial t} + \rho C \mathbf{u} \cdot \nabla T = \nabla \cdot \left(k_{eff} \nabla T \right) + \rho H \frac{\partial f}{\partial t} + Q_{em} + Q_{ext} \tag{8}$$

where $\mathbf{E}$, $\mathbf{B}$, $\mathbf{H}$ and $\mathbf{D}$ are electric field, magnetic induction, magnetic field and displacement current, respectively; and σ is the electrical conductivity. Also, $\mathbf{u}$, T and p are the velocity, temperature and pressure fields; Q_{ext} and F_{ext} are external heat and momentum sources, excluding those from electromagnetic origin. The source terms to the mechanical balance equations are represented by $\mathbf{J} \times \mathbf{B}$ and Q_{em}. The materials properties are given as follows: ρ for density, C for specific heat, μ for viscosity and κ for thermal conductivity. For most practical applications, the magnetic Reynolds number is small, which means the diffusion of the magnetic field is predominant over the convection of it by fluid. Consequently, we have the relation,

$$\mathbf{J} = \sigma \mathbf{E} \tag{9}$$

which suggests that the magnetic field may be calculated as if the liquid were solid. If further one assumes that the material properties are weakly temperature-dependent, then the electromagnetic field can be decoupled from the thermal and fluid flow fields. It is well known that for a wide range of materials processing systems under terrestrial conditions, these approximations are well applicable.

Many numerical schemes now exist that can be applied to solve these complex mathematical equations. Table 1 summarizes the computational methodologies available for the solution of the Maxwell equations and the mechanical (thermal and momentum) balance equations. With a wide spread use of the internet, one is able to search available individual software by an entry of computational electromagnetics and/or computational fluid dynamics and heat transfer. It should be stressed that despite the availability of numerous computational software, either through commercial market or in-house development, an efficient and correct solution of practical problems rests essentially with the understanding of the nature of these problems and the knowledge of the user of the available codes.

Table 1. Comparison of numerical methods for electromagnetics and
thermal fluid problems

Electromagnetics	Thermal fluids

Boundary element method	Boundary element method (limited use)
Finite element method	Finite element method
Discontinuous finite element method	Discontinuous finite element method
Finite difference method	Finite volume method
Volume integral method	

It is noted that the discontinuous finite element method is a new and evolving numerical technique that combines the salient features of the both finite difference and finite elements. The theory of the method and its application to the solution of electromagnetic and mechanical balance equations are documented in a very recent monograph [5].

It is also important to point out that there exists a critical distinction between the solutions of the Maxwell equations and of the mechanical balance equations is that the former in general requires to resolve the distribution of the field over the entire space, or more specifically the boundaries for the electromagnetic field problems are at the infinity. In contrast, the mechanical balance equations often have well defined boundaries and the distribution of these fields is confined within these boundaries.

Multiscale Description of Local Events

As discussed briefly in the introduction, solidification involves spatial and temporal scales in a wide disparity. As seen in Figure 1, the length scales associated with the transition of liquid into solid ranges from atomistic up to the continuum. Any mathematical descriptions need to capture these different scales.

Molecular Dynamics Simulations

The transition of atoms from the liquid to the solid phase is often modeled using molecular dynamics simulations. The calculation is based on the Lagrangian description of individual atoms in motion [6],

$$m\frac{d^2\mathbf{r}}{dt^2} = -\nabla_{\mathbf{r}}\Phi \qquad (9)$$

where $\mathbf{r}$ is the position of an atom and m its mass, and Φ is the intermolecular potential. A widely used model for intermolecular potential is the 6-12 Leonard-Jones potential.

By monitoring the movement of an ensemble of atoms, the atomic structure of the solid-liquid interface can be obtained, which then allows us to determine the parameters characterizing the interface such as interfacial thickness and interfacial surface energy [7]. As solidification is in essence an interfacial phenomena driven by surface energy and interfacial curvature, the interfacial parameters are of crucial importance for the growth of individual crystals.

Phase Field Modeling

The phase field theory is developed as a result of studying nonlinear critical phenomena during phase transitions in superconducting materials and other physical systems [8]. The central idea in phase field modeling of solid-liquid phase transitions is the concept of localized statistical averaging, that is, coarse-graining: which means that some of the microscopic degrees of freedom in a given system are integrated out, leaving an effective system (characterized by an effective free energy or Hamiltonian) with fewer degrees of freedom embodied by the coarse-grained order parameter.

In statistical mechanics, thermodynamic quantities are calculated from the partition function Z, which is related to the free energy F, where $Z = \exp(-F(T)/k_bT)$, and k_b is the Boltzmann constant. By the coarse-grain averaging process, a block is selected and averaging is carried out over a number of molecules in the cell for a microscopic quantity, yielding a block quantity, and then summing over those microscopic quantities that give rise to a given block yields a localized partition function Z', which is related to a coarse-grained free energy F, where $Z' = \exp(-F([\phi], T)/k_bT)$. Here ϕ is the phase field parameter, which marks the change from one phase to another. In this way, Z is a sum of all Z' in the system. The square brackets indicate that $F[\phi]$ is a functional of the phase parameter ϕ. The phase field model for phase transition and interfacial phenomena is concerned about the behavior of the phase field parameter ϕ, which is related to the coarse grain free energy. In general, the free energy $F([\phi], T)$ is a functional of the phase field parameter ϕ and it assumes the following form:

$$F([\phi],T) = \int \left(\tfrac{1}{2}\varepsilon^2 \mid \nabla\phi(\mathbf{r})\mid^2 + f(\phi(\mathbf{r}),T)\right)d\Omega \qquad (10)$$

where ε is the interfacial gradient thickness parameter and $f(\phi(\mathbf{r}), T)$ is a potential function and is a function (but not functional) of $\phi(\mathbf{r})$ and temperature. Because the phase field model is formulated locally and is able to resolve the microscopic phenomena, it has been adopted to model the interfacial phenomena in thermal and fluids systems that are diffusive in nature [4, 43].

In a time-dependent system, an evolution of the phase parameter $\phi(\mathbf{r}, t)$ represents the evolution of the interface. The phase field model is derived based on the fact that the system evolves towards a state that minimizes the free energy,

$$\tau\frac{\partial\phi(\mathbf{r},t)}{\partial t} = -\frac{\delta F}{\delta\phi} + \eta(\mathbf{r},t) \qquad (11)$$

where τ is the relaxation time, and $\eta(\mathbf{r}, t)$ represents the stochastic noise that describes the random effects of the environment, and is averaged to zero over the realization of the noise field $\eta(\mathbf{r}, t)$. For most thermal and fluids applications, this term may be set to zero.

Performing the variation of the functional F, one arrives the following evolution equation for the phase field $\phi(\mathbf{r}, t)$,

$$\tau\frac{\partial\phi}{\partial t} = \varepsilon^2\nabla^2\phi - \frac{\partial f}{\partial\phi} \qquad (12)$$

With f known, this equation can be used to evolve the free surfaces and boundaries and hence the grain growth in solidification processes. The parameters τ and ε are parameters that characterize the interfacial phenomena, which are obtained either through experiments or molecular dynamics simulations.

Celluar Automata

While phase field model is useful for modeling grain growth during solidification, the fine grids required to resolve the local resolution of grain tip curvature can be too expensive computationally for a practical application. An alternative is to use the cellular automata, a numerical method proved to be useful to model the grain growth in both solid deformation processing and solidification processing [9,10,11].

To use the cellular automata technique for a meaningful simulation of grain growth, the individual growth laws must be employed. For some applications, a growth rate based on the assumption of parabolic grain tip has shown a reasonably good result compared with experimental measurements. A more accurate rate expression for grain growth, however, may be obtained through the phase field simulations. We note that the cellular automata is a numerical method, rather than a physical description, for the modeling of local grain growth phenomena during solidification processing.

Multislcale Multiphysics Coupling

Coupling of the multiscales and multiphysics associated with electromagnetically-assisted solidification processing can be a challenging task. Two coupling approaches may be applied. In the first approach, molecular motion may be coupled with microscale phase field models interactively in that an ensemble of molecules, which is taken as apart of the grain-liquid interface, may be used to derive the interfacial parameters. These parameters are then used as input for a phase field model to advance the grain growth. This process is repeated for a new advancement of the interface. The microscale phase field model is then linked with the macroscopic fluid flow and thermal fields, again interactively. This micro-macro coupling requires the selection of multigrids and multi temporal scales. The fluid motion may be influenced by an applied external field, and its effects can be further cascaded onto atomic level. For this type of coupling, an iterative scheme is needed to integrate the atomic calculations all the way to the macroscopic calculations.

For solidification process design and optimization applications, however, the atomic-phase-field-continuum integration, discussed above, may be prohibitively expensive in computation and is impractical, even with the already highly advanced computing platform. A more practical and reasonably well designed first principle approach may be described as follows. For a given solidification problem, molecular dynamics simulations are used to obtain interfacial parameters characterizing the liquid-solid interface. These parameters are then used as input to the phase field model, which is applied to simulate the growth of a single grain and a few grains under a given solidification condition. From these phase field simulations, the rate expression for grain growth is derived, which takes into consideration the complicated factors such as grain impingement and grain orientations. With the growth rate expression so obtained, the cellular automata is then applied to model the grain growth and is coupled with the macroscopic description of the fluid flow, thermal and electromagnetic field distributions. By this approach, the molecular dynamics simulations and phase field simulations are made sequentially and serve the purpose of obtaining the needed rate of growth, while the mutual coupling of the micro-macroscopic phenomena for the simultaneous description of grain growth and thermal, convection and electromagnetic fields is achieved by a combined use of cellular automata and numerical methods for the macroscale field distributions through an easily designed micro-macro spatial grids and micro-macro temporal steps [9-12].

Results And Discussion

In this section, we present some results calculated using the second coupling scheme of multiscale simulations for solidification processing of an aluminum alloy with external electromagnetic stirring.

The system to be studied is shown in Figure 2, where a simple (2-dimensional), idealized, unidirectional solidification system is considered, three sides of which are insulated and cooling is applied on the fourth side. The metal alloy (Al- 4.5%Cu) is initially above its melting point and solidification progresses as the heat is drawn from the right side. The solidifying alloy is stirred by induced electromagnetic forces. These forces are produced by the induction coils through which AC currents are applied in the z-direction, which is perpendicular to the plane of the paper [3]. The mathematical formulations governing these phenomena consist of the Maxwell equations for electrodynamics, the Navier-Stokes equations for fluid flow, the thermal balance equation for heat transfer, which were given in Section 2 above, and nucleation and grain growth calculations.

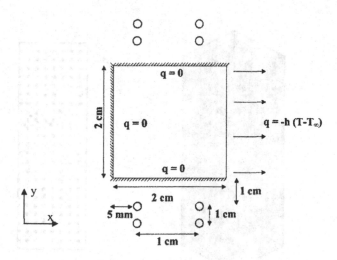

Figure 2. Schematic representation of solidification with induction stirring.

To model the mushy zone effect, the Carman-Kozney equation is incorporated into the F_{ext} term [6]. The boundary conditions for the electromagnetic fields are such that both tangential and normal components of the magnetic and electric fields are continuous at the interface between the air and metal, and they diminish to zero infinitely far away from the coil-metal assembly. At the walls, the no-slip conditions are applied for the velocity.

Molecular Dynamics Simulations

To derive a grain growth rate expression, both the molecular dynamics (MD) simulations and the phase field modeling were applied. An MD model provides solid-liquid interfacial energy and solid-liquid interface thickness – scales on the order of 10^{-9} (m) to order of 10^{-7} (m). We follow the approach presented in [7] and calculate the interfacial parameters using the interfacial fluctuation correlations. As a first approximation, pure aluminum is used, and Figure 3 shows a simple sample molecular system where the liquid-solid interface is resolved at the molecular level.

Phase Field Modeling

Using the interfacial parameters, the phase field model is applied to simulate the single and multigrain growth, from which rate expressions are obtained. A phase field model provides a microstructure evolution at scales on the order of 10^{-7} (m) to 10^{-4} (m). Figure 4 illustrates the phase field modeling results on the evolution of dendritic structures formed during solidification, the typical structure found in ice freezing from water, and the growth of a polycrystal from the melt. The phase field model for our studies for these cases was solved using the discontinuous finite element method, parallelized to enhance the performance for an 8-processor minisupercomputer [13,14]. These results are consistent with those reporeted in literature.

Using the phase field simulations, the rate expression for the grain growth can be derived. Various simulations were conducted and the results obtained confirm with that obtained from empirical data [11].

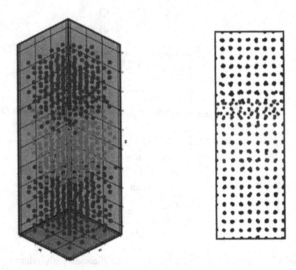

Figure 3. Solid-liquid interface resolved through molecular dynamics simulations.

Coupling of Cellular-Automata with the FE/BE Method

With the grain growth rate expression available, as described above, the Cellular-Automata method is used to model the evolution of the grains during solidification. The algorithm used here is similar to that reported in [9-11] and its implementation in a parallel computing environment was described in detail elsewhere [11]. The algorithm treats nucleation by a Monte-Carlo method and evolves the grain growth by the CA method.

The Maxwell equations and mechanical balance equations are solved using our in-house computational software. A hybrid boundary element and finite element (BE/FE) method is employed to solve the Maxwell equations, which reduces to a single component of vector potential for this problem. The Galerkin finite element method is applied to solve the fluid flow and heat transfer equations and time evolution is advanced with the backward Euler (or implicit) time difference scheme with automatic time step control and the penalty method for pressure approximation [15].

In order to integrate the micro-macro calculations, a multigrid system is used, where macro elements are used for continuum calculations, while fine CA cells are used to evolve the microstructure. A corresponding macro-micro time step is also applied, wherein a macro time-step is employed for heat and fluid flow calculations and a micro time-step for microstructure calculations. For each micro time-step, the undercooling for each CA cell is calculated by interpolating the temperature at the nodes of the element to the center of the CA cell. The solid fraction, for each element, calculated using the CA model, at the end of each macro time-step is used to calculate the latent heat release and the phase interaction effects on fluid flow using Carman-Kozney equation. It should be noted that the micro time-step should be small enough so that the dendrite tip extension never exceeds the cell spacing.

Many simulations were made using the integrated multiscale model presented above. Beucase of the limited space, one set of data is presented below. The computations used 1600 4-node elements and each element was divided into a 50×50 CA grid to give a grid resolution of about 10μm [11]. The parameters used for all computations presented in this paper are given in a thesis, which were confirmed through MD and phase field simulations [11]. It is noted here that the vector potential, fluid flow and temperature fields possess a perfect symmetry with respect to the middle plane, but such symmetry in solidification microstructure is only approximate, although the solidification front itself exhibits perfect symmetry. This is because the detailed grain structure is dependent upon the grain orientation at the micro-scale level that is determined by the local conditions near the wall.

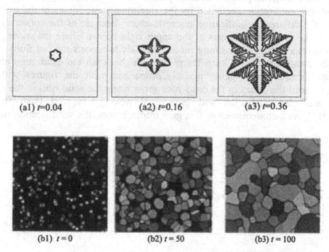

(a1) t=0.04   (a2) t=0.16   (a3) t=0.36

(b1) $t = 0$ (b2) $t = 50$ (b3) $t = 100$

Figure 4. Phase field simulation of grain grwoth: (a1-a3) dendrite growth under adiabatic conditions for various latent heat: $K = 1.6$ is non-dimensionalized latent heat; and (b1-b3) simulation of the impingement and coarsening of multigrain microstructure formed from an undercooled melt [5].

169

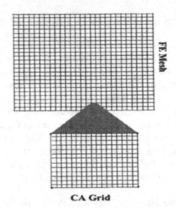

CA Grid

Figure 5. Meshes of finite element and Cellular-Automata for solidification modeling.

Figure 6 depicts the time evolution of fluid flow, temperature distribution and microstructure formation with electromagnetic stirring. Inspection of these snapshots illustrates that the application of the stirring has a profound effect on the solidification behavior of the metal, both in terms of the macroscale solidification front and microscale grain structure. The vigorous stirring results in a flow structure exhibiting two loops rotating in opposite directions and recirculating in the entire pool. The temperature field is distorted by the convective transport in the pool (as evident in the temperature contour plots) and the hot liquid is quickly moved around by the recirculating convection generated by the electromagnetic forces in the melt, leading to the formation of a different solidification microstructure. Because of the convective flows, the solidification of the metal first appears at the upper right corner where the stirring is not fully penetrated (dead-zone area). Also, strong mixing quickly transports the hot fluid to the middle region near the symmetry line to keep temperature higher than the dead zone areas, thereby preventing solidification from taking place. Comparison with the figures without stirring indicates that some of the grains in the dead-zone grow along the solid wall in the early stage of solidification until the nucleation and growth of new grains stop them from growing continuously [11]. As a consequence of this gain impingement, the solidification microstructure is different.

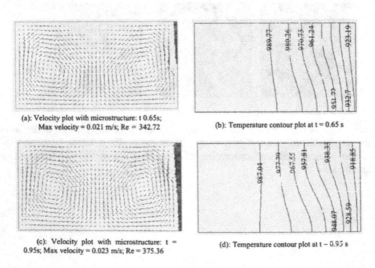

(a): Velocity plot with microstructure: t 0.65s; Max velocity = 0.021 m/s; Re = 342.72

(b): Temperature contour plot at t = 0.65 s

(c): Velocity plot with microstructure: t = 0.95s; Max velocity = 0.023 m/s; Re = 375.36

(d): Temperature contour plot at t – 0.95 s

170

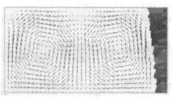

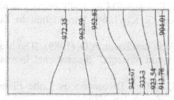

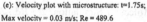

(e): Velocity plot with microstructure: t≈1.75s; (f): Temperature contour plot at t = 1.75 s
Max velocity = 0.03 m/s; Re = 489.6

Figure 6. Dynamic development of electromagnetically driven flows, temperature distributions and microstructure formation in solidifying metals with electromagnetic stirring: $I = 200$ A; $f = 50$ Hz; $h = -3000$ W/m^{2}K; $T_{init} = 1000$ K.

Future Directions

Two important directions in which solidification modeling can be further enhanced. In conducting MD simulations of molecules in motion, the intermolecular potential is required. While the Leonard-Jones pair potential serves a good approximation for many molecular systems, an accurate prediction of the intermolecular potentials requires the solution of quantum mechanics equations. Further development in the multiscale simulation for solidification applications would likely involve the use of quantum mechanics simulations in combination with molecular-phase-field-continuum mechanics simulations.

One other important direction is related to the design of solidification processes and materials. Up to date, simulations using models have been applied to help the design solidification processing systems, whether they are for laboratory testing or for industrial applications. A most effective and efficient way to design solidification processing systems is through inverse design algorithm development. This would represent a paradigm shift. In a inverse design involving multiscale models, the properties of a final solidification product are specified by the customers. The inverse design algorithm will repeatedly use a multiscale model to search for the right materials and processing conditions to meet the requirement of the desired materials properties.

In passing, we should stress that integrated multiscale and multiphysics modeling has opened up an entirely new paradigm for the design of electromagnetically-assisted materials processing systems. A long cherished dream, in which a new material for a given application may be developed by selecting basic chemical elements from the periodic table and is manufactured using appropriate processing conditions, may now be realized by integrated quantum-molecule-phase-field-continuum multiscale multiphysics models.

Acknowledgement

The authors gratefully acknowledge the support of this work by NASA (Grant #: NAG8-1477) and U.S. Army through subcontract from II-VI, Inc.

171

References

[1]   Hoyt, J.J., Asta, M. and Karma, A. (2003) "Atomistic and continuum modeling of dendritic solidification," Materials Science and Engineering R Vol. 41, 121–163.

[2]   Jackson, J. D. (1968) Classical Electrodynamics, Wiley, New York.

[3]   Batchelor, G. K. (1966) Introduction to fluid dynamics, Cambridge University Press, London.

[4]   Voller, V.R. and Brent, A.D. (1989) "The Modeling of Heat, Mass and Solute Transport in Solidification Systems," International journal of Heat Mass Transfer, Vol 32 No 9, pp. 1719-1731.

[5]   Li, B. Q. (2006) Discontinuous Finite Elements in Fluid Dynamics and Heat Transfer, Springer-Verlag, Jan, 2006.

[6]   Fenkel, D. and Smit, B. (2002) Understanding Molecular Simulation, 2^{nD} Ed., Academic Press, New York.

[7]   Su, Y., Asta, M. and Hoyt, J. J. (2004) "Kinetic coefficient of Ni solid-liquid interfaces from molecular-dynamics simulations," Physical Review B Vol. 69, 024108 ~2004.

[8]   Ala-Nissila, T., Majaniemi, S., and Elder, K. (2004) "Phase-Field Modeling of Dynamical Interface, Phenomena in Fluids," in Phase-Field Modeling of Dynamical Interface Phenomena in Fluids, Lect. Notes Phys. 640, 357–388, Springer-Verlag.

[9]   Rappaz, M. and Gandin, Ch.-A. (1993) "Probabilistic Modelling of Microstructure Formation in Solidification Processes," Acta metallurgica et materialia, Vol 41, 345-360.

[10]   Gandin, Ch.-A. and Rappaz, M. (1994) "A Coupled Finite Element-Cellular Automaton Model for the Prediction of Dendritic Grain Structures in Solidification Processes," Acta metallurgica et materialia, Vol 42, 2233-2246.

[11]   Govindaraju, N. (2000) Numerical Simulation of Microstructure Formation During Solidification with Magnetically Driven Flows Using Parallel Computing, MS Thesis, Washington State University, Pullman, WA.

[12]   Li, B.Q. and Anyalebechi, P.N. (1995) "A Micro/Macro Model for Fluid Flow Evolution and Microstructure Formation in Solidification Processes," International journal of Heat Mass Transfer, Vol 38, 2367-2381.

[13]   Shu, Y., Ai, X. and Li, B.Q. (in press) "Phase-field Modeling of Solidification Microstructure using the Discontinuous Finite Element Method," Int. J. Num. Meth. Eng.

[14]   Ai, X., Shu, Y., and Li, B.Q. (in press) "Discontinuous Finite-Element Phase-Field Modeling of Polycrystalline Grain Growth with Convection," Computers and Mathematics with Applications.

[15]   Song, S. P. (1999) Finite element analysis of surface oscillation, fluid flow and heat transfer in magnetically and electrostatically levitated droplets. Ph.D. Thesis, Washington State University, Pullman, WA, 2000.

Materials Processing under the Influence of External Fields
Edited by Qingyou Han, Gerard Ludtka, Qijie Zhai
TMS (The Minerals, Metals & Materials Society), 2007

COUPLED MODELING FOR ELECTROMAGNETIC SOLIDIFICATION TRANSPORT PROCESSES OF ALLOY CASTINGS

Daming Xu, Yunfeng Bai, Jichun Xiong, Hengzhi Fu
School of Materials Science and Engineering, Harbin Institute of Technology
E-mail address: damingxu@hit.edu.cn

Keywords: EM-materials processing, Solidification, Transport phenomena, Computer Modeling

Abstract

In applications of Electromagnetic (EM) fields to advanced liquid→solid (L/S) materials processing, the EM-fields exert significant influences on the heat, mass and momentum transport behaviors in the solidifying materials *via* Joule heating/Lorentz forces. It is of importance to develop modeling techniques to simulate the L/S EM-processing of alloy products. This article describes the recent work at HIT on the modeling for EM-processing of TiAl castings/ingots, using authors' proposed/extended continuum model and the solution approaches to the dendritic solidification transport phenomena under an arbitrary EM-field. Future related-tasks are suggested.

Introduction

The techniques of electromagnetic processing of materials(EPM) [1] have got more and more important and wider applications to advanced materials processing in the past decades, e.g., using a static or traveling magnetic field to damp (electromagnetic brakes(EMBR)) the alloy melt convections in crystal growth and continuous casting (or DC casting) processes of various non-metallic and metallic materials [2-4]; applying various alternating (harmonic) electromagnetic(EM)-fields to DC ingot casting for the meniscus shapes and surface qualities controls [5, 6]; and grain refining of the solidifying ingots by an EM-stirring/vibrating technique *etc* [7-9].

In such a liquid→solid(L/S-) EPM process, the applied EM-fields, as an external field, will exert great influences on the heat, mass and momentum transport behaviors of the solidifying materials *via* Joule heating and/or Lorentz forces. On the other hand, there are usually many complex influencing factors and technological parameters involved in an L/S-EPM process. Therefore, it is of both the theoretical and practical interests to develop computer modeling techniques to simulate the EM-solidification transport phenomena (STP) in an L/S-EPM process. The aim of this paper is to present the recent work at HIT on the computer modeling for EM-STP behaviors of TiAl castings/ingots under an arbitrary EM-field, using the authors' proposed/extended continuum model and the numerical solution methods. Future related-tasks in this area are suggested.

EM-STP-Based Model

The EM-STP model for the heat energy, species mass and momentum of liquid flow in a solidifying alloy casting/ingot under an arbitrary EM-field is extended from an author's previously proposed continuum model [10-13] based on the theory of Thermo-Magneto Hydro Dynamics. The model may take the following formulation form with the assumptions made in Ref.[13]:

Heat Transfer Equation

$$(\rho c_P)_m \frac{\partial T}{\partial t} + \nabla(f_L \rho_L c_{PL} V_L T) = \nabla(\lambda_m \nabla T) + \rho_S h \frac{\partial f_S}{\partial t} + q_J \qquad (1)$$

Where $q_J = J_G^2/\sigma$ is the induced Joule heat.

Species Mass Transfer Equation

$$\frac{\partial(\rho C)_m}{\partial t} + \nabla(f_L \rho_L V_L C_L) = \nabla\big[D_L \nabla(f_L \rho_L C_L)\big] + \nabla\big[D_S \nabla(f_S \rho_S C_S)\big] \qquad (2)$$

Alloy Solidification Characteristic Function

$$T_{Liq} = F\left(C_L^*\right) \qquad (3)$$

Mass Conservation Equation

$$\frac{\partial \rho_m}{\partial t} = -\nabla(f_L \rho_L V_L) \qquad (4)$$

Momentum Transfer of Bulk/Interdendritic Liquid

$$\frac{\partial(f_L \rho_L V_L)}{\partial t} + \nabla\big[(f_L \rho_L V_L)V_L\big] = \nabla\big[\eta \nabla(f_L V_L)\big] - \nabla(f_L P_L) - \frac{\eta f_L^2}{K}V_L + f_L \rho_L g + F_L \qquad (5)$$

Where, the EM-body (Lorentz) force on moving liquid in the continuum model is expressed as:

$$F_L = \sigma f_L\left(E + V_L \times B\right) \times B = f_L\left\{J_G \times B + \sigma\big[(V_L \cdot B)B - (B \cdot B)V_L\big]\right\} \qquad (6)$$

The present continuum model comprises seven independent physical fields: temperature T, liquid volume fraction f_L ($f_S = 1 - f_L$), concentration C_L, pressure P, flow velocity V_L, magnetic flux density B and the induced current density J_G. The EM-fields in a whole solidification system, including the domains of alloy casting, mold, cooler and the environmental air etc, are described by Maxwell's equations and the corresponding constitutive relationships [13].

In the authors' more recent modeling for 2D harmonic EM-STP problems, the solution of Maxwell's equations is converted to solve the following PDE for the vector magnetic potential A, by defining $B = \nabla \times A$ [14]:

$$\frac{\partial^2 \dot{A}_z}{\partial x^2} + \frac{\partial^2 \dot{A}_z}{\partial y^2} = -\frac{\dot{J}_z}{v} + j\frac{\omega\sigma}{v}\dot{A}_z \qquad (7)$$

174

Numerical Solution Approaches to the Model Couplings

For a general dendrite solidification case with any incomplete SBD, the time-differential mixture-averaged composition (TDMAC) term in the macroscopic Equation (2) is closely related to the microscale solute redistributions in the dendrites, and can be expressed as the following fully differential TDMAC form [11]:

$$\frac{\partial(\rho C)_m}{\partial t} = (\rho_S C_S)^{\bullet} \frac{\partial f_S}{\partial t} + \Phi f_S \frac{\partial(\rho_S C_S)^{\bullet}}{\partial t} + (\rho_L C_L) \frac{\partial f_L}{\partial t} + f_L \frac{\partial(\rho_L C_L)}{\partial t} \qquad (8)$$

Correspondingly, the time-differential mixture-averaged (TDMAD) density term in the solidification mass conservation Equation (4) can be approximately expressed as [12]:

$$\frac{\partial \rho_m}{\partial t} \approx \rho_S^{\bullet} \frac{\partial f_S}{\partial t} + \Phi f_S \frac{\partial \rho_S^{\bullet}}{\partial t} + \rho_L \frac{\partial f_L}{\partial t} + f_L \frac{\partial \rho_L}{\partial t} \qquad (9)$$

In equations (8) and (9), the functions for the unified microscale-parameter Φ, accounting for any finite SBD effects in the solidifying dendrite phases, can be expressed as [11]:

$$\Phi = \frac{\theta \varphi}{1 + \theta \varphi}, where: \theta = \frac{(1 + \beta)kf_S}{f_L^2}; \varphi = \frac{D_S(T)}{R_f} \zeta \cdot A_{2N} \qquad (10)$$

There exist several close correlations and couplings in the proposed EM-STP model of equations (1)-(10). Table I gives a summary for the model solution methods proposed by the authors for the major couplings/correlations involved.

Table I. Major Couplings/Correlations Involved in the present EM-STP Model and the Solution Approaches Proposed by the Authors

Couplings/Correlations	Solution Approaches Proposed	References
Species mass transfer in the ingot&castings' and dendrites' length scales (Macro-/Micro-Segregation)	A unified microscale parameter, Φ, is proposed for the species mass transfer Eq.(2), which has a function form of Eq.(10) for any SBD and dendrite morphologies	[11,12,15]
Correlations between the solidification heat/mass and the momentum transfers	Numerically separated by a Up-Wind scheme	[16-19]
T-f_S-C_L coupling in alloy solidification processes	Solved by a simultaneously numerical procedure proposed and extended with Φ	[16,18]
P-V_L coupling in the alloy solidification processes	Solved by a direct-SIMPLE algorithm proposed and extended to EM-STP problems	[17,19]
EM–STP couplings	One-way EM-influences on STP, under the assumptions of unvaried boundaries and EM-properties of the solidifying materials	[13]
EM-FEM/STP-FDM(FVM) Numerical couplings	Treated with the authors' FEM–FDM(FVM) data converting technique/computer codes	[20]

175

Example Computations

Feasibility and computational efficiency of the proposed EM-STP continuum model and the solution methods have been shown in published papers [11-13, 15-21]. In this article, the availabilities and the recent progress in the authors' EM-STP coupled modeling are further presented by the following two groups of sample calculations, using the modeling methods described in the previous sections and a model alloy of pseudo-binary γ(TiAl)-55at.%Al system.

The first group of computations is on directional solidification (DS) processes of a 2D blade-like casting under a harmonic EM-field. The EM-DS configuration and the used numerical techniques for the FEM/FVM-coupled EM-STP simulations are similar to that described in Refs. [13] and [20]. The major progress in the authors' coupled modeling is that the EM-fields are calculated by an authors' recently developed FEM-based computer codes [14]. This is a necessary step towards to a fully (two-way) EM-STP coupled modeling, especially for a case with melt free-surface movements/oscillations.

Figure 1 shows the FEM-calculated B-vectors in the EM-DS system by the authors' program and the compared results by commercial software of ANSYS 8.0. It shows that the two sets of the results, both for the entire calculation domains and in the EM-DS core regions, are closely similar. The authors' FEM-EM fields are coupled to the FVM-based STP simulations using the same techniques of Ref. [20]. Figure 2 shows the coupled EM-STP simulation results compared with those only under the gravity. From Figure 2 it can be seen that, under the harmonic EM-fields as shown by Figure 1, the convection in bulk liquid region is greatly enhanced, which in turn leads to several negative pressure holes to form, and the contours of temperature and concentration towards to a more stronger convection-transfer type.

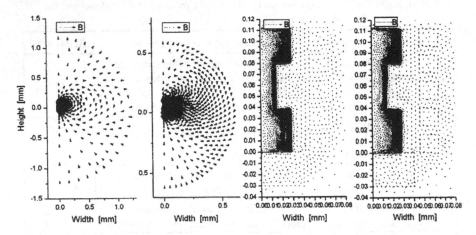

B-vectors in the whole EM-DS region B-vectors around the blade-like casting/coils/cooler

Figure 1   FEM-calculated B-vectors in an EM-directional solidification system (right half) of TiAl blade-like castings—Comparisons with that of ANSYS 8.0 FEM-calculations (each right).

176

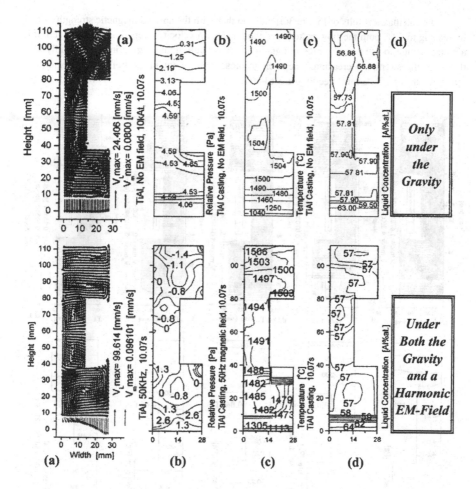

Figure 2 FEM/FVM coupled-simulation results of EM-STP behaviors at a directional solidification time of 10 sec.: (a) velocity vectors of liquid phases; (b) contours of relative pressure in the liquid; (c) contours of temperature; (d) contours of liquid concentration.

In the second example group, in stead of the harmonic EM-fields, transverse static magnetic fields [13, 19] of different strengths up to 50T are applied to the DS system to investigate EMBR effects of ultra strong magnetic on the STP behaviors in the blade-like casting DS processes. A comparison of the calculated results is shown in Figure 3.

It can be seen from Figure3 that, the relative strong natural convections in the bulk liquid region of the directionally solidifying casting can be effectively suppressed by a 0.066T magnetic field, except that driven by the volume-contraction (mainly solidification shrinkage). The presently calculated results show that such volume–contraction-driven liquid flow can not be stopped even under a very strong magnetic field of 50T, see Figure 3(a)-50T.

The calculated results of Figure 3(a) indicate that with the applied magnetic strength further increasing to 50T, both the flow pattern and the maximum velocity keep almost unchanged. It is also interesting to see that, the pressure gradients in the bulk liquid region increase rapidly with the magnetic strength increasing. The average pressure gradient in the bulk liquid region can reach a value of 28kPa/mm! See Figure 3(b)-50T.

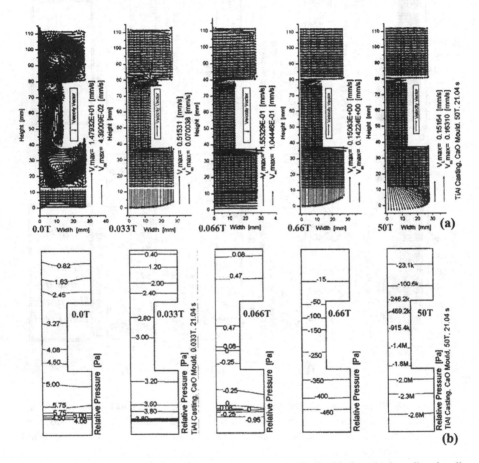

Figure 3 Velocity vectors(a) and relative pressure contours in liquid phase(b) in a directionally solidified γ(TiAl)-55at.%Al shaped castings at t=21.0 sec under gravity and with an applied transverse static magnetic field of different strengths: 0.0T, 0.033T, 0.066T, 0.66T and 50T.

Remarks on the Tasks Needed to be Further Carried Out

The present model and the corresponding solution methods have been further shown to be successful for a coupled EM-STP numerical simulation. However, there still exist several important aspects in the EM-STP-based computer modeling, which need to further work on due

178

to the complicated natures of the practical engineering problems. These may be summarized as the follows.

1. Multicomponent/multiphase issue: Practical engineering alloys are almost multicomponent systems [21]. To track the correct solidification path on a liquidus surface and the possible variant eutectic valleys/peritectic paths etc in a multicomponent solidification process, it is a great challenge to extend the proposed binary T-f_S-C_L numerical solution [16, 18] to an arbitrary T-Σf_{Si}-ΣC_{Lj} coupling case combined with a piece of Calphad software, e.g. ThermoCalc [22].

2. 3D-STP applications: It has been shown that a critical issue in a 3D-STP computer modeling is to successfully and efficiently solve the liquid phase-fraction -associated pressure-increment (LPAPI)-matrix for the 3D P-V_L coupling [23].

3. Turbulent flow: With a strong EM-stirring or/and heavy alloy melt inlet, the fluid flow is highly possible in a turbulent manner. This may need to incorporate a k-ε or k-ω two-equation turbulent model [24, 25] in the proposed EM-STP continuum modeling.

4. Melt free-surface movements/oscillations: This will cause full EM-STP correlations between the fields of [B, J] and [T, V_L, V_S, P, f_{Si}, C_{Lj} etc]. Techniques of numerical boundary curvature/movement tracking, as accurately and efficiently as possible, are needed [26, 27].

5. Correlations with the solidification structures: The multiple length-scales Φ-parameter modeling [11, 12, 18, 15] has, in fact, accounted for the influences of microscale morphology of the solidifying phases (dendrite/interdendritic eutectic etc) on the macroscale species mass transport behaviors to some extents. It is well known that, however, the solidifying phase movements on the castings/ingots length-scales, in a multiphase flow with the melt convection, will exert much more significant impacts on the macroscale distributions of solidification defects, including the segregations, inclusions and porosity etc. The key-tasks in solving such correlations include the multi-length-scale modeling for the cellular-to-equiaxied transaction (CET) criteria (alloy types/compositions-, solidification parameters- and flow behaviors-dependent), and for the nucleation ratio functions (also alloy/mold-materials types-, melt structures/qualities- and flow behaviors-dependent) etc.

6. Towards a more general continuum modeling for segregation/porosity/inclusion formations in multicomponent/multiphase-flow systems [28]. It is believed that, the further modeling tasks on the related multiple-scale parameters and on the efficient numerical algorithms (practical computationally acceptable) for all the complicated correlations/couplings are much more challenging than just to propose these mathematical models themselves.

Acknowledgement

Supports from a Key China NNSF of grant No. 50291012 to parts of the present work are acknowledged.

References

1. S. Asai, "Recent Development and Prospect of Electromagnetic Processing of Materials", *Science and Techn. of Advanced Materials*, 1 (2000), 191-200.

2. K. Kakimoto and H. Ozoe, "Oxygen Distribution at a Solid-Liquid Interface of Silicon under Transverse Magnetic Fields", *Journal of Crystal Growth*, 212 (2000), 429-437.

179

3. H.L. Yang et al., "Mathematical Study on EMBR in a Slab Continuous Casting Process", *Scandinavian Journal of Metallurgy*, 27 (1998), 196-204.
4. N. Genma et al., "The Linear-Motor Type In-model Electromagnetic Stirring Technique for the Slab Continuous Caster", *ISIJ Int.*, 29 (12) (1989), 1056-1062.
5. J.W. Evans, "The use of electromagnetic casting for Al alloys and other metals", *Journal of Metals (JOM)*, 1995 May: 38-41.
6. T. Li et al., "Study of Meniscus Behavior and Surface Properties during Casting in a High-Frequency Magnetic Field", *Metal. Mater. Trans. B*, 26B (1995), 353-359.
7. Ch. Vivès, "Effects of Forced Electromagnetic Vibrations during the Solidification of aluminum Alloys: Part II. Solidification in the Presence of Colinear Variable and Stationary Magnetic Fields", *Metal. Mater. Trans. B*, 27B (1996), 457-464.
8. K. Sugiura and K. Iwai, "Effect of Operating Parameters of an Electromagnetic Refining Process on the Solidified Structure", *ISIJ International*, 44 (11) (2004), 1410-1415.
9. E. Beaugnon et al., "Material Processing in High Static Magnetic Field. A Review of an Experimental Study on Levitation, Phase Separation, Convection and Texturation", *Journal de physique I*, 3 (2) (1993), 399-421.
10. Daming Xu, "A Continuum Model for Macrosegregation Formations in Ingots Solidification" (Ph.D. thesis, Harbin Institute of technology, 1989), 13-39.
11. Daming Xu, "A Unified Micro-Scale Parameter Approach to Solidification -Transport Phenomena-Based Macrosegregation Modeling for Dendritic Solidification: Part I. Mixture Average Based Analysis", *Metal. Mater. Trans. B*, 32B (2001), 1129-1141.
12. Daming Xu, "A Unified Micro-Scale Parameter Approach to Solidification -Transport Process-Based Macrosegregation Modeling for Dendritic Solidification: Part II. Numerical Example Computations, *Metal. Mater. Trans. B*, 33B (2002), 451-463.
13. Daming Xu et al., "Heat, Mass and Momentum Transport Behaviors in Directionally Solidifying Blade-like Castings in Different Electromagnetic Fields Described Using a Continuum Model", *Int. J. Heat and Mass Transfer*, 48 (2005), 2219-2232.
14. Xiong Jichun, "FEM-Based Software for Predicting the Electromagnetic Fields in Ti Alloys EM-Directional Solidification System and EM-STP Coupled Simulation" (Master thesis, Harbin Institute of technology, 2006), 15-33.
15. Daming Xu et al., "Influences of Dendrite Morphologies and Solid-Back Diffusion on Macrosegregation in Directionally Solidified Blade-Like Casting", *Materials Science and Engineering A*, 344 (1-2) (2003) 64-73.
16. Daming Xu and Qingchun Li, "Numerical Method for Solution of Strongly Coupled Binary Alloy Solidification Problems", *Numerical Heat Transfer, Part A*, 20 (1991), 181-201.
17. Daming Xu and Qingchun Li, "Gravity- and Solidification-Shrinkage-Induced Liquid Flow in a Horizontally Solidified Alloy Ingot", *Numerical Heat Transfer, Part A*, 20 (1991), 203-221.
18. Xu Daming et al., "Numerical Solution to $T-\varphi_S-C_L$ Coupling in Binary Dendrite Solidification with Any Solid-Back Diffusion Effects", *Trans. of Nonferrous Metals Society of China*, 13 (5) (2003), 1149-1154.
19. Daming Xu et al., "Numerical Simulation of Heat, Mass and Momentum Transport Behaviors in Directionally Solidifying Alloy Castings under Electromagnetic Fields Using an Extended Direct-SIMPLE Scheme", *Int. J. for Numerical Methods in Fluids*, 46 (2004), 767-791.
20. Yunfeng Bai and Daming Xu et al., "FEM/FDM-Joint Simulation for Transport Phenomena

in Directional Solidifying Shaped TiAl Casting under Electromagnetic Field", *ISIJ International*, 44 (7) (2004), 1173-1179.

21. Daming Xu et al., "Numerical Simulation of Transport Phenomena in Solidification of Multicomponent Ingot Using a Continuum Model", *J. of Materials Science and Technology*, 17 (1) (2001), 67-68.

22. Thermo-Calc Software AB, TCC^{TM} *Thermo-Calc Software User' Guide, Version Q* (Stockholm, Sweden: Stockholm Technology Park, 2004).

23. Daming Xu and Jun Ni, "A Numerical Approach of Direct-SIMPLE Deduced Pressure Equations to Simulations of Transport Phenomena during Shaped Casting", *Proc. of the Int. Multi-Symposium on Computer and Computational Sciences*, vol.2, ed. J. Ni and J. Dongarra et al. (IEEE Computer Society, Hangzhou, 20-24 June 2006), 815-821.

24. W.P. Jones and B.E. Launder, "The Prediction of Laminarization with a Two-Equation Model of Turbulence", *Int. J. Heat and Mass Transfer*, 15 (1972), 301-314.

25. Xiaoqing Zheng et al, "Turbulent Transition Simulation Using the k-ω Model", *Int. J. for Numerical Methods in Engineering*, Vol.42, 1998, pp.907-926.

26. W. Shyy et al., *Computational Fluid Dynamics with Moving Boundaries* (Washington, DC: Taylor&Francis Pub., 1996), 1-19.

27. Ruxun Liu, and Zhifeng Wang, *Numerical Simulation Methods and Moving Boundary Tracking* (Hefei: Pub. of China Science and Technology University, 2001).

28. Daming Xu et al., "A Continuum Model for Ingot Solidification Transport Phenomena and Defects Formation", *The Chinese Journal of Nonferrous Metal*, 7 (Suppl.1) (1997), 259-262.

Materials Processing under the Influence of External Fields
Edited by Qingyou Han, Gerard Ludtka, Qijie Zhai
TMS (The Minerals, Metals & Materials Society), 2007

EFFECT OF COMPLEX ELECTROMAGNETIC STIRRING ON SOLIDIFICATION STRUCTURE AND SEGREGATION OF 60Si 2CrVAT SPRING STEEL BLOOM

Weidong Du[1], Kai Wang[2], Huigai Li[1], Xufeng Liu[1], Qijie Zhai[1], Pei Zhao[3]

[1] Center for Advanced Solidification Technology(CAST), Shanghai University, Shanghai , 200072, P.R. China
[2] Baoshan Iron & Steel Co., Ltd. Special Steel Branch Manufacturing Management Dept, Shanghai , 200940, China
[3] Central Iron and Steel Research Institute, Beijing 100081, China

Keywords: Complex electromagnetic stirring, Spring steel, Solidification structure, Segregation.

Abstract

The effect of complex electromagnetic stirring combined M-EMS with F-EMS on solidification structure and segregation in high strength 60Si2CrVAT spring steel bloom has been studied. It is found that the extent of the equiaxed structure is considerably increased and solidification structure is modified by the complex electromagnetic stirring. Moreover, the ratio of central carbon segregation remarkably reduced. Therefore, the problem of inner porosity and composition segregation can be solved with the process in spring steel 60Si2CrVAT bloom.

Introduction

Electromagnetic stirring of steel in continuous casting bloom is a widespread technique for improving the quality of cast products in industry. To obtain a better quality of high strength spring steel bloom, complex electromagnetic stirring which combines stirring the liquid steel in the mould (M-EMS) with stirring in the final solidification zone (F-EMS) is carried out at Baoshan Iron & Steel Co., Ltd. Special Steel Branch Industry. Although there is a great deal of research work[1-7] about effects of rotary electromagnetic field on metal solidification structure and segregation, its relevance to industrial complex electromagnetic stirring[8] remains limited.

The aim of the present work is twofold. First, we want to obtain a better understanding on the solidification structure of high strength 60Si2CrVAT spring steel bloom of the combined effects of M-EMS and F-EMS. For this purpose the macrostructure of continuous casting blooms based on experimental results for 60Si2CrVAT spring steel is studied. Second, we want to ascertain the effects of complex electromagnetic stirring on the homogeneity of carbon in the continuous casting bloom of 60Si2CrVAT spring steel. The results of this study are presented in this paper.

Materials and Experimental Method

Plant trail

A kind of high strength spring steel has been assessed for the complex EMS. The compositions are showed in Table I. To assess the effects of complex EMS on the quality of bloom, a plant trial is carried on Concast S22-10.25CCS arc bloom caster with shielded casting practice.

Operative conditions of the caster machine are briefly listed in Table II. The main parameters of complex electromagnetic stirring are listed in Table III. The process technology is used in the following process: 100tDC-100tLF-100tVD-CC.

Table I. Main chemical composition of steels (wt.%)									
Grade	C	Si	Cr	Mn	V	Ni	P	S	Cu
60Si2CrVAT	0.58	1.5	1.17	0.6	0.15	0.12	0.01	0.003	≤0.15

Table II. Operative conditions of the caster machine at Baoshan Iron & Steel Co., Ltd. Special Steel Branch Industry.	
Steel superheating (K)	~30
Arch radius (mm)	10250
Nozzle from tundish to mold	SEN
Bloom dimensions (mm × mm)	220×220
Mold length (mm)	800
Mold taper	curve
Electromagnetic stirring	M-EMS+F-EMS
secondary cooling zone	3 segments
Casting speed (m/min)	0.9
Strand	5

Table III. Electromagnetic stirring parameters		
Parameter	M-EMS	F-EMS
Current (A)	300	150
Frequency (Hz)	5	11
Rotation direction	No change	10s alternate change

Macrograph of bloom section

To assess the quality of bloom, 20mm length section blooms are intercepted at the strand start, middle and bottom locations. Moreover a 10mm×20mm×110mm special sample is intercepted at the section bloom along the centerline. The macrograph and defects are revealed by hot acid erosion (50wt%HCl+50wt%water, 333k).

Carbon segregation analysis experiments

184

The slices on transverse sections are cut by Wire Electron Discharge Machine (WEDM). The positions are shown in Figure1. Each slice is polished and washed by alcohol. The mass of each slice is about 0.5~1.0g. Then the carbon contents are measured by Leco CS600 Carbon & Sulfur Determinator.

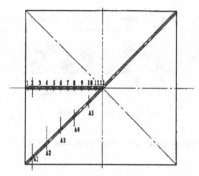

Figure 1. The sketch of slice location in 60Si2CrVAT bloom section for carbon analysis.

Results and discussions

Quality of blooms and solidification structure

The macrographs of 60Si2CrVAT cast with complex EMS are presented in Figure 2 and Figure 3. At the transverse sections center of the blooms, where the last location solidifies, the shrinkage cavities are all small. In the transverse section, fine equiaxed structure predominates. The central shrinkage porosity of blooms is very fine and is complemented with uniform distribution from top to bottom of strand. There are no inner crack and corner crack in all transverse sections.

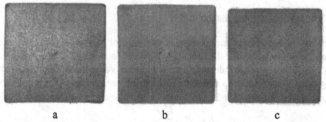

a   b   c

Figure 2. Macrograph of the in transverse sections of spring steel 60SiCrVAT bloom (a) top of the strand, (b) middle of the strand, (c) bottom of the strand strands.

The sketch of the ordinary cast solidification macrostructure is shown in Figure 4, with the surface chill zone, the intermediate columnar zone, and the central equiaxed zone. All three of these zones are seen in continuous casting blooms of plain carbon or low-alloy steel. However, as shown in Figure3, there are four zones in spring steel 60Si2CrVAT bloom, the surface chill zone, the intermediate columnar zone, the transition zone from columnar to equiaxed zone, and the central equiaxed zone. We can determine the length of four zones in the bloom section. The area of equiaxed zone is about 25% of the bloom section area, and the transition zone is about

10% of the bloom section area. The central porosity and shrinkage cavity with the complex EMS are reduced, and no clear white band is found in the bloom section. All of above proves that the complex EMS process used in plant trial can obtain the markedly increased extent of the equiaxed zone and better quality bloom.

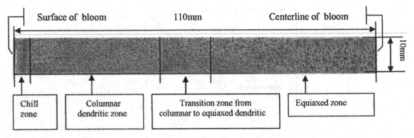

Figure 3. Macrostructure of sample of spring steel 60Si2CrVAT bloom by complex EMS.

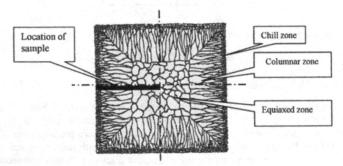

Figure 4. Sketch of ordinary cast solidification macrostructure.[9]

Centerline carbon segregation and mechanism

The centerline carbon segregation is expressed as the ratio of carbon analysis of the centerline C_0 to the averaged carbon analysis of the cast C_B. The ratio is referred to as a carbon segregation coefficient K_C. σ is the standard deviation of the mean value of the segregation coefficient. Carbon segregation in the bloom is characterized by the mean value of the carbon segregation coefficient $\overline{K_c}$. This value is obtained based on the carbon analysis of a sample. Thus

$$K_C = \frac{C_0}{C_B} \tag{1}$$

$$\overline{K_c} = \frac{\sum K_{C1} + K_{C2} \cdots + K_{Cn}}{n} \tag{2}$$

C_B is the carbon analysis of a tundish sample, its value equal to 0.577. $\overline{K_c}$ is the mean value of centerline carbon segregation coefficients. K_{C1} to K_{Cn} is the individual segregation coefficients. n is the number of the individual K_C.

The spring steel 60SiCrVAT carbon segregation profiles on centerline and on diagonal line of the bloom transverse section are shown in Figure 5. An average centerline carbon segregation of $\overline{K}_c = 0.986$, with $\sigma = 0.048$ is obtained with complex EMS for continuous casting. Moreover, an average carbon segregation of $\overline{K}_c = 0.987$, with $\sigma = 0.025$ on diagonal line of the bloom transverse section is also obtained. As seen from these data, the centerline carbon segregation is reduced significantly in the spring steel 60SiCrVAT with complex EMS.

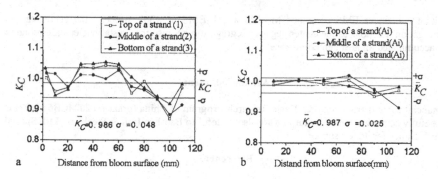

Figure 5. Carbon segregation on transverse section of 60Si2CrVAT bloom of (a) on centerline and (b) on diagonal line.

It is observed from Figure 5 (a) and (b) that negative segregation appears in the central zone. Central segregation depends on the movement of liquid and crystal grains in final solidification. Continuous casting bloom with complex EMS presents negative central segregation. It can be explained by crystal grain detachment theory presented by Ohno[10]. During continuous casting, lots of free grains come into existence at nozzle and mold walls, and a great amount of solid particles come into existence under meniscus. These solid particles are flushed and concentrated at the center of solidification front edge by stirring spiral vortex which induced negative segregation. At the same time, because the solidification front does not advance smoothly, large equiaxed dendrite crystals tend to form partial bridge. If solute can sufficiently flow under electromagnetic stirring force in the final solidification, the solution in the mushy zone would be homogenized in mush zone. Particles easily accumulated at the partial bridge. Therefore the central zone presents negative segregation.

Figure 3 vs Figure 5(a) reveals the relationship between spring steel 60SiCrVAT solidification structure and carbon segregation. As we can see, at chill zone about 3-5 mm in Figure 3, solute cannot reject out from chilling crystals in time, so experiment data show positive segregation. At the beginning of columnar crystals growth, carbon content of the crystals is low. With the growth of the columnar crystals in columnar dendritic zone, carbon solute in the solid-liquid interface gradually accumulates, reaching the climax in the transition zone from columnar to equiaxed dendritic. At the final solidification, force convection arises typically from suction caused by electromagnetic stirring force, and free crystals in bulk deposit. Therefore the central equiaxed zone presents negative segregation. At the area close centerline carbon content increase again, because the solute rejected at the end of solidification.

Conclusions

187

When the complex EMS combined M-EMS and F-EMS techniques is used in the spring steel 60Si2CrVAT continuous casting process, the bloom solidification structure forms 4 zones: chill equiaxed dendrite zone, columnar dendrite zone, columnar & equiaxed dendrite zone and center equiaxed dendrite zone. With low superheat temperature and stirring of E-EMS and M-EMS, the center equiaxed dendrite zone is extended, the solidification structure is modified and the forming centerline shrinkages decreased in casting process.

With the comples EMS combined M-EMS and F-EMS in whole process, the center carbon segregation functions as the center negative segregation because the crystal nucleus are prone to be accumulated in near centerline zone.

Acknowledgment

Financial support from National Basic Research Program of China (grant no.2004CB619107) is gratefully acknowledged. The authors are also grateful to Baoshan Iron & Steel Co., Ltd. Special Steel Branch for its test support.

References

1. B. Willers et al., "The columnar-to-equiaxed transition in Pb-Sn alloys affected by electromagnetically driven convection," *Materials Science and Engineering A*, 402 (2005), 55-65.
2. Carmen Stelian et al., "Solute segregation in directional solidification of GaInSb concentrated alloys under alternating magnetic fields," *Journal of Crystal Growth*, 266 (2004), 207-215.
3. L.B. Trindade et al., "Numerial Model of Electromagnetic Stirring for Continuous Casting Billets," *IEEE Trans. Magn.*, 38 (6) (2002), 3658-3660.
4. K. Fujisaki et al., "Magneto Hydrodynamic Calculation for Electromagnetic Stirring of Molten Metal," *IEEE Tran. Magn.*, 34 (4) (1998), 2120-2122.
5. R. Hirayama and K. Fujisaki, "Combined System of AC and DC Electromagnetic Field for Stabilized Flow in Continuous Casting," *IEEE Trans. Magn.*, 41 (10) (2005), 4042-4044.
6. S. Sato and K. Fujisaki, "Responsiveness of Free Surface Flow to M-EMS Exiting Frequency," *IEEE Trans. Magn.*, 41 (10) (2005), 3451-3453.
7. D.-G. Zhou et al, "Effect of CC Process Parameters on Carbon Segregation of Bearing Steel," *Iron and Steel* (China) 21(1999), 131-134.
8. J.-C. Liet al ., "Effect of Complex Electromagnetic Stirring on Inner Quality of High Carbon Steel Bloom" *Materials Science and Engineering A*. 425 (2006), 201-204.
9. Merton C. Flemings. *Solidification Processing* (New York, NY: McGraw-Hill, Inc.,1974), 135.
10. Ohno A. *Metal solidification- Theory, Practice and Application*, trans. J.D. Xing (Beijing China: Engineering industry press, 1986), 91-100.

Materials Processing under the Influence of External Fields
Edited by Qingyou Han, Gerard Ludtka, Qijie Zhai
TMS (The Minerals, Metals & Materials Society), 2007

INFLUENCE OF ELECTRIC CURRENT PULSE ON SOLIDIFICATION

STRUCTURE OF HYPOEUTECTIC AL-CU ALLOY

Xi-Liang Liao[1,2] Yu-Lai Gao[1], Ren-Xing Li[1], Qi-Jie Zhai[1]

[1] Center for Advanced Solidification Technology (CAST), Shanghai University, Shanghai, 200072, P.R. China;
[2] School of Mechanical Engineering, Shandong University, Jinan 250061, P.R. China

Keywords: Electric current pulse, Al-Cu alloy, Solidification structure, Refining mechanism

Abstract

The aim of this paper is to investigate influence of electric current on solidification structure of hypoeutectic Al-Cu alloy. The results show that using electric current pulse remarkably refined the macrostructure of Al-Cu alloy. Moreover, with increase of the current intensity, not only the grain size decreases, but also the primary a-Al phase of microstructure was transformed from dendritic crystals into equiaxed ones. In addition, it was also found that with the increase of solute content, the refining effect of electric current pulse on macrostructure was weakened. The refinement mechanism of electric current pulse is that nucleus can be broken off from the mold wall and migrate into the melt by means of pulse electromagnetic force.

Introduction

Al-Cu alloy is extensively used in aerospace and some important mechanical parts because of excellent heat-resistance and mechanical properties,. But its application is largely limited due to wide crystallization temperature range, which can result in casting defects, such as hot cracking. It is well known that defect formation is closely related to the solidification process and structure, therefore, improving the solidification process and solidification structure of Al-Cu alloys contribute to solve these problems. Addition of the trace elements and rare Earth elements can refine solidification structure of Al-Cu alloys[1-4], but would simultaneity contaminate alloy too. In recent years, some external fields, such as electric current pulse(ECP), magnetic field and ultrasonic[5-7], are applied to solidification process, and results show that it can markedly refine solidification structure of metal[5-7]. ECP treatment is a new contamination-free method to refining solidification structure and easy operation.[8,9]. In this paper, the effect and main mechanism of ECP on refining solidification structure of hypoeutectic Al-Cu alloy was investigated.

Experimental

The experimental apparatus was presented in Fig.1, which included an electric current pulse generator with the capability of 1500μf, and a set of temperature measurement. The pulse power supply frequency is in the range of 100-1000Hz, and the electric voltage is in the range of 12.5 to 3000V.

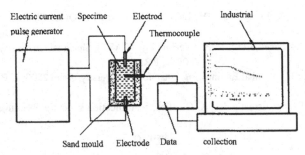

Fig.1. Schematic sketch of the experimental setup

The hypoeutectic Al-Cu alloys were prepared by pure Al(99.7%) and pure Cu(99.99%) through intermediate frequency induction furnace. Alloy was melted in an electric resistance furnace to 150K above melting point, held for 30 min, and then was poured into two sand moulds at the same time. In the same condition, one cast was solidified with ECP, and the other was solidified without ECP. A NiCr-NiSi thermocouple was used to record temperature of the melt at middle of the melt. The data acquisition system consisted of an A/D converter and a computer. In the same cooling condition, one cast was without ECP, and another was with ECP until the end of solidification.. The sample size was φ40mm×160mm, which were cut along longitude and transverse cross-sectioned at the midline, polished and etched for metallographic examination. The etching reagent was the mixture of nitric acid, hydrofluoric acid, hydrochloric acid, and water in the ration of 3:3:9:5.

Experimental Results

The macrostructure of Al-6.64%Cu alloys with different current intensity is shown in Fig.2. Fig.2(a) is the solidification structure without ECP. The solidification structures on ECP treatment of different current intensity are presented in Fig.2(b)-(e). Results indicate that with the increase of ECP, the average grain sizes gradually decrease. The average grain size of samples with different current intensity is shown in Fig.3. It is can be seen that grains decrease from 1.764mm without ECP to 0.792mm with 8000k_l A ECP. The microstructure of Al-6.64%Cu alloys with different current intensity is shown in Fig.4. Fig.4(a) reveals that in the sample without ECP, primary phase a-Al is coarse dendrite crystal with longer primary arm and developed secondary arm. When ECP is applied to samples, column-equiaxed crystal transfer takes place for the microstructure and secondary-arm gradually disappears.

190

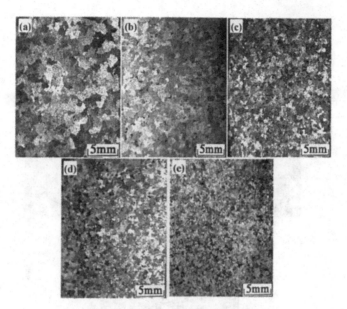

Fig.2. The effect of ECP on macrostructure of Al-6.64%Cu alloys
(a)Untreated,(b)3200k_1 A,(c)4800 k_1 A,(d)6400 k_1 A,(e)8000 k_1 A

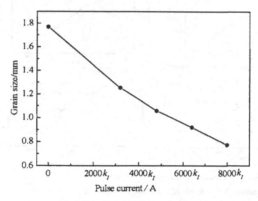

Fig.3. The effect of ECP on grain size

191

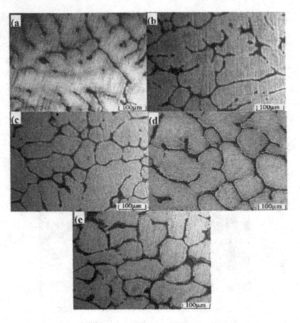

Fig.4. The effect of ECP on microstructure of Al-6.64wt%Cu alloys

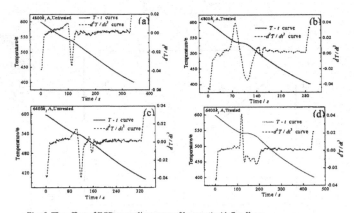

Fig. 5. The effect of ECP on cooling curve of hypoeutic Al-Cu alloys
(a) 4800k$_i$ Untreated, (b) 4800k$_i$ treated, (c) 6400k$_i$ Untreated, (d) 6400k$_i$ treated

Discussion

According to Maxwell equations ($\nabla \times H = J, \nabla \times E = -\frac{\partial B}{\partial t}, \nabla \Box B = 0, \nabla \Box J = 0$), when ECP is imposed on

192

the molten alloy, the induced magnetic field being perpendicular to current direction is produced by ECP, which can give rise to Lorentz force along the radial direction. Since the molten metal is exposed to radial press force, strong convection formation in the melt. When the melt is pour mould, nucleation of primary a-Al crystals is usually generated on the mould wall firstly. The metal liquid scours the mould wall continuously due to convection induced by ECP. Thus, a-Al crystals can be broken up into pieces from the mold wall and migrate into the melt. This course occurs continuously until the formation of the solidification shell on mould wall, which make the nucleation rate of a-Al crystals remarkably increase. Moreover, melt convection also enhances heat exchange between the center melt and the melt near the mould wall. Thus formation time of solidified shell near the mould wall increases, which is benefit to increase nucleation rate and decrease temperature gradient ahead of growing crystal. Consequently, growth of dendrite crystal will be suppressed. According to results of temperature measurement, applying ECP can prolong solidification time and delay formation of solidified shell on the mould wall, which contribute to promote refinement of solidification structure. Fig.5 shows the cooling curve of samples with different current intensity. It indicates that ECP makes eutectic reaction time of samples becomes longer. As a result, solidification time increases. The increase of solidification time benefits to form nucleus, Fig.6 shows the grain size vs Cu content of Al-Cu alloys with and without ECP. It shows that with the increase of Cu content, the refining effect become weaker in the same ECP. According to phase diagram, the crystallization temperature range of hypoeutectic Al-Cu alloys decreases with the increase of Cu content, which shortens the nucleation time of primary phase a-Al on mould wall. Thus, the nucleus number induced decrease by ECP.

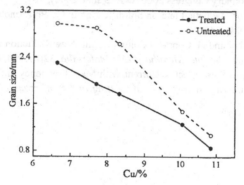

Fig.6. The effect of Cu content of Al-Cu alloys on grain size with ECP

Conclusions

The macrostructure of Al-Cu alloy is remarkably refined by the application of electric current pulse. With the increase of current intensity, the grain size of decrease, and primary a-Al phase transforms from dendritic crystals to equiaxed ones. In addition, with the increase of Cu content, the refining effect of ECP on macrostructure is weakened.

Acknowledgements

The authors gratefully acknowledge the financial supports from the Pre-research Foundation of the National Basic Research Program (973 Program, grant No. 2004CCA07000) and the Science and Technology Committee of Shanghai Municipality (Grant No. 04XD14008).

Reference

1. S. Q. Dong et al, "The Effects of Sb and Bi on the Hot Cracking Tendency of Casting Al-Cu Alloys," *J. XI'AN Institute of Technology*(in Chinese), 22(2) (2002), 159-163.
2. G. f. XU et al. "Effects of Micro-Addition of Sc on Microstructure and Properties of Al$_2$Cu Alloy," *Transactions of Materials and Heat Treatment*(in Chinese), 25(2)(2004), 15-18.
3. X. W. Wei and M. Zeng, "The Effects of Rare Earth Elements on The Fluidity And Tendency on Hot Craking of Al$_2$Cu Cast Alloy," *Foundry Technology*(in Chinese), (3)(1997), 46-48.
4. J. W. Mao, T. N. JIN and G.-F. XU, "Microstructure of Al-Cu Alloys Containing Scandium,"

   The Chinese Journal of Nonferrous Metals(in Chinese), 15(6)(2005), 923 - 928.

5. F. Li, L.L.Regel, and W.R.Wilcox, "The Influence of Electric Current Pulses on the Microstructure of the Mn/Bi Eutectic," *Journal of Crystal Growth*, 223(2001), 251 – 264.
6. Q. M. Liu et al, "Effect of Ultrasonic Vibration on Structure Refinement of Metals," (Paper Presented at the World Foundry Congress 2006. U.K. 5 June 2006), 167.
7. H. Yasuda et al, "Effect of Magnetic Field on Solidification in Cu-Pb Monotectic Alloy," *ISIJ International*, 43(2003): 942-949.
8. J. P. Barnak, A. F. Sprecher and H. Conrad, "Colony (grain) Size Reduction in Eutectic Pb-Sn Castings by Electropulsing," *Scripta. Metallurgica et Materialia*, 32(6)1994, 879-884.
9. Y. FANG et al. "Effects of Pulse Electric Current with Main Frequency and Low Voltage on Grain Refinement of Pure Aluminum," *Special Casting and Nonferrous Alloys*(in Chinese), 26(3)2006,141-143.

PULSED ELECTROMAGNETIC FORMING AND JOINING
Sergey Golovashchenko,

Ford Motor Company
MD3135, 2101 Village Road; Dearborn, MI 48124, USA

Keywords: Sheet Metal Forming, Coil, Springback.

Abstract

In this paper some options of how electromagnetic forming (EMF) can assist to expand the capabilities of conventional forming and joining technologies are discussed. Three different areas where EMF has potential for significant expansion of capabilities of traditional technologies are reviewed: 1) restrike operation to fill sharp corners of automotive panels; 2) low energy method of springback calibration; 3)joining of closed frames with an openable coil. Each of these applications was demonstrated in laboratory conditions and the description of the tooling is provided in the paper. An efficient technique of fabricating the flat coil from a flat plate by using water jetting technology enables a cost effective coil design, which can be reinforced by a system of nonconductive bars. Technology of low-energy calibration of stamped parts was demonstrated for calibration of U-channels made of aluminum alloys, mild steels and advanced high-strength steels.

Introduction

In pulsed electromagnetic forming and joining, a coil can be described as a tooling replacing a punch in conventional forming operations. Depending upon the coil configuration, various stamping processes can be carried out, such as tube expansion, tube compression and flat sheet forming [1]. However, due to a number of issues including coil durability, maximum stamping rate, safety implications etc., electromagnetic stamping technologies have not been able to compete with conventional stamping on the press if conventional processes provided the desirable result. Therefore, our strategy is to use electromagnetic forming (EMF) to expand the capabilities of conventional forming and joining technologies. In this paper we will discuss three different areas where EMF has potential for significant expansion of capabilities of traditional technologies: 1) restrike operation to fill sharp corners of automotive panels; 2) low energy method of springback calibration; 3) joining of closed frames with an openable coil.

Restrike operation of parts preformed with conventional stamping technology

Due to the tendency that electromagnetic pressure is usually rapidly decreasing as soon as the blank moves away from the coil, deep drawing operations are not feasible for EMF, unless some special arrangements are made to move the coil close to already deformed blank. Another option is to use several steps of EMF with coil configuration at the following step corresponding to the blank shape after the previous step. This approach may be acceptable for low-volume production. However, for high-volume it may not be appropriate. More attractive is the combination of conventional forming and EMF, when EMF is applied as a restrike operation for difficult-to-form areas. It has been proven experimentally [2] that for corner filling operation EMF can produce significant improvement in formability due to a number of effects described in [3], including high strain rate, bending-unbending, significant hydrostatic pressure and coining effect. Some complication of EMF employment for restrike operation is dictated by limited space for the coil. Typically, to have high-efficiency EMF process, the ratio of the inductance of the coil-with blank related to the total inductance of the EMF machine-coil-blank system should be approximately 0.8...0.9 [1]:

$$(1) \quad K_1 = \frac{L_{coil+blank}}{L_{mach} + L_{connection} + L_{coil+blank}} \approx 0.8...0.9.$$

In other words, the inductance of the coil should dominate over the inductance of the EMF machine and its connection to the coil. To satisfy this requirement, the coil should have multiple turns, since the inductance of the coil can be described in a following way

$$(2) \quad L_{coil+blank} \sim n^2,$$

where n is the number of turns in the coil. Multiple turns are difficult to accommodate in limited space for restrike operation and also may not be feasible from the coil strength and durability point of view. Evidently, the cross-section of the turn dictates its stiffness against torsion moment developed as a result of interaction of adjacent turns of the coil and also due to the skin effect. Usage of a single turn coil usually leads to low efficiency of the process. Suggested solution is in a system including a flat coil and a concentrator [4] combining high efficiency of the multiturn coil and structural strength of a single-turn coil. The important feature of the flat coil is that its spiral is relatively easy to fabricate: it can be manufactured from a flat plate [5] using water jetting technology, which provides significant cost reduction compared to end milling technology. If compared to a winding technique, the water jetting technology provides more accurate shape of the surface and allows maintaining a uniform and rather small clearance between the coil and the blank, since during the winding procedure originally straight bar gets curved as a result of plastic bending, producing distortion of the original cross-section. Also, in suggested new design, the coil should be reinforced to balance significant forces in between the turns of the coil. This can be accomplished by employing non-conductive rods as it is indicated in Figure 1. Expansion of the coil can be prevented by putting the coil into a metallic bandage [5].

The design of such a reinforced coil was developed based upon the analysis of failure modes of the coils in our laboratory practice. The insulation between the turns was fabricated from micarta sheet. The thickness of the sheet was specified to be equal to the clearance between the turns of the coil produced by water jetting technology.

Spiral — Insulation plates

Reinforcing rods

Figure 1: Flat multiturn coil employed for laboratory experiments

The design of the laboratory flat concentratorwas applied to a corner filling operation. A flat blank of aluminum alloy 6111-T4 was used as a sample. It was positioned on the surface of a die fabricated to demonstrate the process in laboratory conditions. The tooling, employed in this experiment is shown in Figure 2.

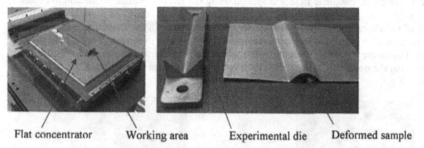

Flat concentrator   Working area   Experimental die   Deformed sample

Figure 2: Flat coil with flat concentrator employed for laboratory experiments of corner filling operation.

Springback calibration.

The idea of springback calibration using high-speed forming technologies has been known for several decades. In many cases, it was applied for calibration of cylindrical shells bent and welded from a flat sheet. An example of such a process is shown in Fig.3 , where the welded blank is inserted inside the calibration die. The expansion coil is positioned inside the blank. During high voltage discharge of capacitors, internal pressure pulse is applied to the blank. Driven by this pressure, the blank accelerates within the clearance between the blank and the die and then impacts the die. Typically, the calibration effect is accomplished due to radial stresses, generated during the high-rate impact of the blank with the die. Any source of pulsed pressure, such as EMF, electrohydraulic or explosive forming would be appropriate to provide such calibration mechanism. In many cases it requires to accelerate the blank up to the speed of 100-200 m/sec.

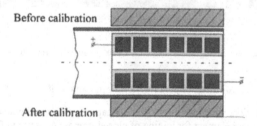

Before calibration

After calibration

Figure 3. Schematic of pulsed electromagnetic calibration of welded blanks

In our recent development, we suggested another calibration process based upon elimination of internal stresses in the blank during propagation of induced electric current [5]. After conventional stamping operation, springback of the blank can be calibrated by deforming the blank elastically back to the desirable shape and then running high voltage discharge through the coil positioned in close vicinity to the targeted zones for calibration. These targeted zones are areas of the blank with significant internal stresses, created during clamping of the blank to its designated shape. Such process requires significantly less energy compared to traditional pulsed calibration technology. It was discussed in details for a simple test of flattening pre-bent sheet in [7]. Similar technology was demonstrated for calibration of U-channels with flat coil-flat concentrator system. Suggested process was verified for aluminum alloys AA6111-T4 and AA5754 and also for steels BH210, DP500 and DP600. The laboratory tooling is shown in Figure 4. The electromagnetic field was applied using the flat coil, similar to the coil in Figure 1, and a flat concentrator which produced electromagnetic field in the area of curved areas at the bottom of a U-channel. In order to generate the internal stresses in the area of radii at the bottom of the U-channel, a simplified approach was used: the blank was deformed elastically by two C-clamps. The elastic character of such a loading procedure was verified by loading-unloading test and checking the geometry of the blank. The examples of samples from different materials are shown in Fig.5.

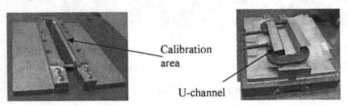

Calibration area

U-channel

Figure 4: Flat coil with flat concentrator employed for U-channel calibration

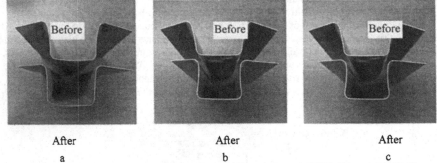

After   After   After

a   b   c

Figure 5: Comparison of U-channels before and after calibration: a- DP500, 0.65mm ; b- BH210, 0.7mm; c- AA6111-T4 - 0.98mm

Joining of parts with an openable coil

Finally, the third application, which we discuss in this paper is joining with an openable coil. Openable coil is a necessary element for joining of closed frames. Assembled frame can not be removed from a traditional cylindrical coil [1]. Evidently, the requirements for joints of structural parts are very high in terms of both strength and durability. Therefore, the desired joint should include high strength crimp or cold weld, which can be produced using EMF. In this paper we describe the design of an openable coil, which consists of two independent miltiturn coils connected in series [6], each of which is mounted in its own shell. The shell was made of electrical insulation material with electrical connections to an EMF machine. On the side surface of each coil, there is a concave work zone with the shape of a tubular component. The coils are positioned in such a way that their work zones are against each other. Together both work zones form a closed-loop work cavity for the tubular component. The coils are connected in series by electrically conductive plate. The first turn of the first coil is connected to one pole of the EMF machine and the last turn of the other coil is connected to the other pole of the EMF machine. In this case, two coils act independently, but their combination is equivalent to a single cylindrical coil. The schematic of the openable coil is illustrated in Figure 6. In order to increase efficiency of the coil, special grooves, making the path for electric current more preferable through the work surface rather than the opposite surface of the coil, were machined. The pictures of the actual coil validated in laboratory conditions are shown in Figure 7. This coil was employed for joining a cylindrical tube to a cylindrical mandrel and also a rectangular tube to an inner rectangular mandrel, both shown in Figure 8. In the second case, an openable concentrator was involved in the joining process [6]. In both cases, the cylindrical and rectangular mandrels had grooves which were filled with tubes' material.

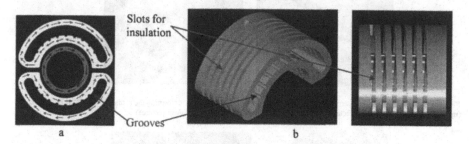

Figure 6: Schematic of an openable coil for radial compression of tubes: a – schematic of the current flow; b – view of one spiral

The assembled openable coil is shown in Fig.7. It is capable of radial compressing of the tubes with the external diameter, corresponding to the internal diameter of the coil. Radial clearance between the blank and the working surface of the coil is usually 1 mm. In order to expand the range of possible operations with described openable coil, openable concentrators (Fig.8) with the dimensions, corresponding to the dimensions of compressed tubes, were fabricated. The examples of joints assembled with the described openable coil are shown in Figure 9. The tubes were fabricated of aluminum alloy 6061. The round tube had 70mm outer diameter and was 3.2mm thick. The rectangular tube had outer dimensions of 38.1mm x 25.4 mm, and had the wall thickness of 3.2mm.

Conclusions

1. Forming of local areas can be conducted using a multiturn coil and a flat concentrator. The multiturn coil was fabricated from a flat plate using water jetting technology and inserting insulation parts cut out from micarta flat sheet.

2. Low energy process of springback calibration has been developed based upon the idea of elimination of elastic residual stresses in stamped blank. The process was demonstrated for calibration of U-channels from aluminum alloys and high-strength steel.

3. Openable coil, comprising of two independent coils connected in series, has been demonstrated for joining of round and rectangular tubes.

Openable concentrator

Electric connection between the
halves of the coil

Figure 7: An openable coil in assembled (left) and disassembled (right) position: deforming the round tube with the coil spiral (upper pictures); deforming the rectangular tube with an openable concentrator (lower pictures)

Figure 8: Examples of joints assembled using the openable coil

References

1. Igor Beliy, Saul Fertik, Lev Khimenko, Electromagnetic Metal Forming Handbook . (Kharkov, USSR:Visha Shkola, 1977)
2. S.Golovashchenko, V.Mamutov, V.Dmitriev, A.Sherman. "Formability of sheet metal with pulsed electromagnetic and electrohydraulic technologies," Proceedings of TMS symposium "Aluminum-2003," San-Diego, 2003, p.99-110.
3. J.Imbert, S.Winkler, M.Worswick, D.Olivera, S.Golovashchenko."The effect of tool/sheet interaction in damage evolution of electromagnetic forming of aluminum alloy sheet. Transactions ASME. Journal of Engineering Materials and Technology, 2005, N1, p.145-153.
4. S.Golovashchenko, "Apparatus for Electromagnetic Forming a Workpiece," US Patent Application # 11/163411. October 18, 2005.
5. S.Golovashchenko, V.Dmitriev, A.Krause, P.Canfield, C.Maranville, "Apparatus for Electromagnetic Forming with Durability and Efficiency Enhancements, US Patent Application # 10/967978, October 19, 2004.
6. S.Golovashchenko, V.Dmitriev, A. Sherman "An Apparatus for Electromagnetic Forming, Joining and Welding." US Patent 6,875,964 B2, April 5, 2005.
7. S.Golovashchenko "Springback calibration using pulsed electromagnetic field," Proceedings of NUMISHEET'2005, Detroit, 2005, p.284-285.

Materials Processing under the Influence of External Fields
Edited by Qingyou Han, Gerard Ludtka, Qijie Zhai
TMS (The Minerals, Metals & Materials Society), 2007

SOLIDIFICATION BEHAVIOUR UNDER INTENSIVE FORCED CONVECTION

Z. Fan, G. Liu, M. Hitchcock and A. Das

BCAST (Brunel Centre for advanced Solidification Technology)
Brunel University, Uxbridge, Middlesex UB8 3PH, UK

Keywords: Solidification, nucleation, convection, atom cluster

Abstract

So far solidification of an alloy with specified compositions has largely remained as a natural process with little control over both nucleation and growth processes, producing a typical growth-controlled microstructure, which is often coarse, non-uniform and with severe chemical segregation. To address this limitation, we have recently developed a dynamic approach to solidification control, in which intensive forced convection (intensive shear stress-strain field) is applied to the solidifying liquid through a twin-screw device. Under such conditions, nucleation becomes the dominant process, creating a nucleation-controlled microstructure, which is fine and uniform throughout the entire cast body. Our research on this subject has led to the discovery of a number of new solidification phenomena, such as continuous nucleation, effective volume nucleation, spherical growth, physical undercooling and physical grain refinement. In this paper, we offer an overview of our work on solidification under intensive forced convection and present a theoretical framework for solidification under external physical fields.

Introduction

Solidification plays a very important role in materials science and technology. It provides the starting point for microstructural evolution during subsequent thermal mechanical processing, and dictates the final physical and mechanical properties of all the engineering components in the cast form. Although substantial progress has been made in the past few decades on growth of crystalline phases from the parent liquid phase [1], our understanding of the nucleation process is still very limited [2]. Nucleation has so far remained largely a natural process for a given alloy composition, with very little external interference. In addition, solidification research is mainly focused on growth (to a much less degree on nucleation) under static (or quiescent) conditions, and solidification under dynamic conditions has been more or less ignored [1]. As a consequence, nearly all the solidification processing technologies produce growth-controlled microstructures in the as cast condition, which consists of a chilled zone at the surface, a coarse equiaxed zone at the centre and a columnar zone in between, and this microstructural heterogeneity is compounded by the large degree of chemical segregation at both macro and micro levels [3]. In order to achieve fine and uniform microstructure and uniform chemistry, extensive solid-state deformation processing has to be used, which is proven to be inefficient and costly. To change such situation, solidification research should be focused on understanding the nucleation process and developing mechanisms for nucleation control, so that a fine and uniform microstructure (nucleation-controlled microstructure) can be produced in the as-cast state without relying heavily on solid state deformation processing.

In recent years, we at BCAST have developed a dynamic approach to nucleation control utilising a high shear twin-screw device, in which solidification takes place under intensive forced convection with high shear rate and high intensity of turbulence. In this paper we briefly summarise our results on solidification behaviour under intensive forced convection, in terms of continuous nucleation, effective volume nucleation, spherical growth, physical undercooling and physical grain refinement.

The high shear twin-screw device

During the development of a semisolid casting process (Rheo-diecasting, RDC for short) [4], we have created a twin-screw mechanism for delivering high shear rate and high intensity of turbulence to the solidifying liquid metal (Figure 1). The twin-screw device consists of a barrel and a pair of screws. The barrel is equipped with both heating elements and cooling channels for accurate temperature control. The screws have specially designed profiles to achieve functions, such as co-rotating, fully intermeshing and self-wiping. The fluid flow in the running twin-screw device is characterised by high shear rate, high intensity of turbulence and rapid heat extraction (not necessarily high cooling rate) [5]. The shear rate can be several orders of magnitude higher than magneto-hydrodynamic stirring. Although it is difficult to quantify the degree of turbulence inside the twin-screw device, we understand that the fluid flow is always turbulent even if the screws are running at low rotating speed. As a consequence of such fluid flow characteristics, solidification takes place in the twin-screw device under the following unique conditions [6]:

- Uniform temperature field, no temperature gradient in the liquid;
- Uniform chemical composition field, no concentration gradient in the liquid;
- Well dispersed nucleation agents throughout the entire volume of the liquid;
- Fast heat extraction, thus absence of recalescence.

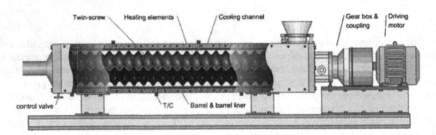

Figure 1. Schematic illustration of the twin-screw high shear device. The barrel is equipped with heating elements and cooling channels for temperature control, the screws are co-rotating, fully intermeshing and self-wiping. It provides high shear rate, high intensity of turbulence and fast heat extraction.

Such unique conditions for solidification make the twin-screw device a powerful tool for solidification research under controlled dynamic conditions. A number of new solidification phenomena have been discovered in the past few years, some of which will be presented briefly in the following sections.

Continuous nucleation

According to the classical nucleation theory [7], nucleation in nearly all the casting processes is a one-off event. Once the superheated liquid metal is poured and in touch with the relatively cold mould surface, "Big Bang" nucleation takes place, and numerous nuclei are created [8]. However, the majority of those nuclei are dissolved into the oncoming superheated liquid, and

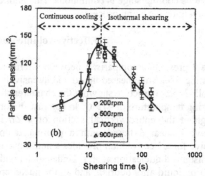

only a small proportion of nuclei in the vicinity of the mould surface can survive. Such survived nuclei will dictate the subsequent solidification process to create the columnar and coarse equiaxed zones, resulting in a growth-controlled microstructure, which is usually coarse, non-uniform and with severe chemical segregation.

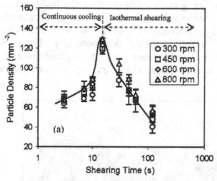

Figure 2. Particle density of the primary phase as a function of shearing time during solidification under intensive forced convection inside the twin-screw device. The results indicate that nucleation under intensive forced convection is continuous. (a) A357 Al-alloy; (b) AZ91 Mg-alloy.

Under the dynamic conditions provided by the twin-screw device the nucleation behaviour is expected to be vastly different. As has been described elsewhere [6], solidification in the twin-screw device is divided into two distinctive stages; continuous cooling between the pouring temperature and the semisolid temperature set by the barrel temperature, and the isothermal shearing at the semisolid temperature. Figure 2 presents the particle density of 2 different alloys as a function of shearing time during solidification inside the twin-screw device. The particle density increases with the shearing time at the continuous cooling stage, but decreases at the isothermal shearing stage. One should bear in mind that the solid fraction increases with time at the continuous cooling stage, remains constant at the isothermal shearing stage, and that the particle size of the primary phase is more or less constant (35-40µm) during the continuous cooling stage [9]. The increase in particle density implies that the nucleation process at the continuous cooling stage is continuous. This is in direct contrast to the "Big Bang" nucleation in the conventional casting processes.

Effective volume nucleation

Figure 3 presents a comparison between microstructures of two Zn ingots produced by gravity diecasting from superheated liquid and intensively sheared liquid. Figure 3a is a typical growth-controlled microstructure showing the characteristic 3 zones produced by gravity die casting with a pouring temperature of 475°C. In contrast, Figure 3b shows a fine and uniform microstructure throughout the entire vertical section of a Zn ingot produced by gravity die casting of an intensively sheared liquid with a pouring temperature of 420°C. As discussed previously, intensive forced convection provides a liquid metal with low superheat, uniform temperature and well-dispersed nucleation agents. Under such conditions, nucleation would take place throughout the entire liquid after pouring, and all the nuclei created would survive and contribute to the final solidified microstructure. We refer to this as effective volume nucleation [6]. Please note that in these experiments shearing was applied to the liquid metal at temperatures above the liquidus, and that nucleation occurred in the mould without shearing. Therefore, nucleation in such cases was not continuous.

After nucleation, numerous nuclei will compete for growth by consuming the liquid around them. As we understand that the initial growth is always spherical, and the spherical growth will continue until the instability of the solid/liquid interface occurs [10]. Due to the effective nucleation, solidification may have completed before reaching this instability. Therefore, effective volume nucleation usually results in fine and equiaxed grains throughout the entire casting without dendritic growth. We refer to such microstructures as nucleation-controlled microstructures.

(a) (b)

Figure 3. Microstructures of pure Zn ingots produced by gravity diecasting. (a) Direct pouring of liquid Zn at 475°C; (b) continuous cooling from 475°C to 420°C under intensive forced convection inside the twin-screw device before pouring. A mild steel mould, with a diameter of 30mm, at room temperature was used for the experiments.

Effective volume nucleation was found to be operational in the rheo-diecasting (RDC) process, at both the primary solidification stage in the twin-screw slurry maker, and the secondary solidification stage inside the die cavity [6]. The RDC components have a fine and uniform microstructure throughout the entire component regardless of the section thickness [11,12]. In addition, the concept of effective volume nucleation has been applied to develop the rheoforming technologies for processing wrought Al- and Mg-alloys [13].

Spherical growth

In conventional casting processes, solidification occurs under quiescent condition. Due to the existence of a temperature gradient and chemical heterogeneity in the liquid phase, the primary phase will usually grow dendritically under constitutional undercooling, resulting in a coarse, non-uniform microstructure with severe chemical segregation. The recent Monte Carlo simulations of solidification under intensive forced convection [14] revealed that the growth morphology changes from dendritical under pure diffusional field, to rosette under laminar flow conditions, and to spherical growth under turbulent flow conditions (Figure 4). In addition, the simulation results also indicated that the growth rate of the primary phase increases with the increase of shearing intensity. Under laminar flow conditions, a floating dendrite grows rapidly not only at the tip of dendrite arms, but also laterally around the dendrite arms, resulting in a rosette morphology. However, under turbulent flow conditions, liquid metal can penetrate the inter-dendritic arm region, transport the rejected solute atoms rapidly into the bulk liquid, and eliminate the constitutional undercooling. Consequently, the interface stability increases, favouring spherical growth.

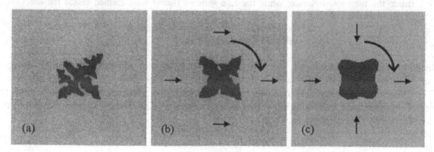

Figure 4. Monte Carlo simulation results showing the effects of convection on the growth morphology. (a) Diffusional field; (b) laminar flow; (c) turbulent flow. The growth morphology changes from dendritic under diffusional field, to rosette under laminar flow and to spherical under turbulent flow.

Spherical growth under intensive forced convection predicted by Monte Carlo simulation has been verified by experiments with the RDC process for different alloys. Figure 5 shows the microstructure of A357 Al-alloy produced by the RDC process. It has been found that solidification under intensive forced convection always produces the primary phase with similar particle size, regardless of alloy composition. For instance, the particle size is about 40μm for Mg-alloys [15], 45μm for Sn-Pb based alloys [16] and 50μm for Al-based alloys [11]. A plausible explanation to the independence of primary particle size on alloy composition is that constitutional undercooling does not exist under intensive forced convection.

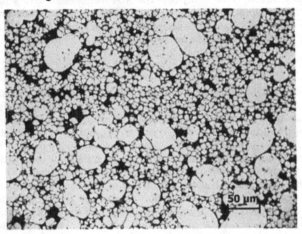

Figure 5. Microstructure of A 357 Al-alloy produced by intensive forced convection in the RDC process. The large spheres (45μm) were formed under intensive forced convection inside the twin-screw device, while the fine spheres (6μm) were formed inside the die cavity under high cooling rate.

Physical undercooling

Recently we have discovered the Non-Newtonian behaviour of metallic liquids using a concentric cylinder viscometer with high temperature and high shear rate capability (Figure 6) [17]. Generally, the isothermal viscosity of liquid metals and alloys increases linearly with the increase of shear rate, exhibiting shear thickening behaviour. This non-Newtonian behaviour can be described by the following equation:

$$\eta = \eta_o + \rho r^2 f(\upsilon)\dot{\gamma} \tag{1}$$

where ρ is the bulk mass density of the liquid metal, η_o is the viscosity of liquid metal when shear rate is approaching 0, $\dot{\gamma}$ is the shear rate, and $f(\upsilon)$ is a tensor-valued function of atomic packing density (υ) in liquid metals. In the $\eta-\dot{\gamma}$ plot (Figure 6), η_o is the intercept viscosity and $\rho r^2 f(\upsilon)$ is the slope. Since viscosity is the most structure sensitive physical property of fluids, the non-Newtonian behaviour implies that the atomic arrangement in liquid metals can be manipulated by an intensive shear stress-strain field. In terms of atom clusters in the liquids, the hypothesis is that intensive shearing can reduce the cluster size.

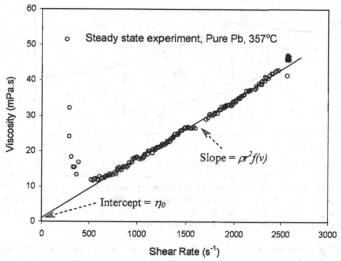

Figure 6. Steady state viscosity of pure Pb at 357°C as a function of shear rate obtained by the shear rate ramping experiment. The time duration for the linear increase of shear rate from 0 to 2600/s was 1 minute.

Figure 7 illustrates schematically the atom cluster size (r) as a function of temperature (T) under different conditions. Under equilibrium conditions (without shearing), the atom cluster size increases with decreasing temperature. Nucleation takes place at a temperature, T_N, where r reaches the critical cluster size for nucleation (r^*). ΔT (T_M-T_N) is the undercooling required for nucleation, which is usually very small. Under a shear stress-strain field, r is reduced, and the $r(T)$ curve shifts downwards. Nucleation can only take place at a lower temperature (T_{NP}). The undercooling ΔT_P (T_M-T_{NP}) is achieved by a physical field, and is thus referred to as 'physical undercooling' to distinguish it from, a) thermal undercooling due to high cooling rate, and b) constitutional undercooling due to the compositional gradient. Physical undercooling will enhance nucleation rate, leading to a finer microstructure. If the shearing intensity increases beyond a certain level, the $r(T)$ curve will not intercept the $r^*(T)$ curve until an extremely high undercooling, and nucleation will be practically eleminated, instead, the undercooled liquid will undergo a glass transition to produce amorphous materials. Figure 7 provides a dynamic approach to the nucleation control.

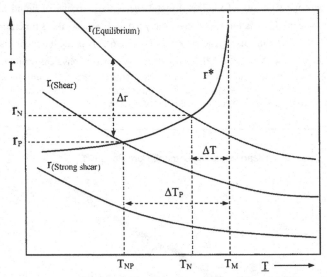

Figure 7. Atom cluster size (r) as a function of temperature (T) under different intensity of shearing to illustrate the dynamic approach to nucleation control. Intensive forced convection will reduce the atomic cluster size from the equilibrium position ($r_{equilibrium}$) to a new dynamic equilibrium condition (r_{shear}) sustained by the high shear, resulting in an increased physical undercooling (ΔT_P). T_M is the melting temperature, ΔT is the undercooling required for nucleation without shearing, and r^* is the critical nucleus radii.

The concept of physical undercooling has been confirmed by our recent experimental results. Figure 8 shows the physical undercooling of pure Zn (industrial purity) as a function of shear rate obtained by continuous cooling of liquid Zn in the twin-screw device. The physical undercooling of pure Zn increases linearly with the increase of shear rate.

Physical grain refining

Physical grain refining is a concept we introduced to describe the effect of intensive forced convection on the refinement of solidified microstructure. Generally, the intensive forced convection is caused by a external physical field (shear stress-strain field in this case) applied to the melt during solidification.

To achieve grain refinement, the conventional wisdom is to add a chemical agent (grain refiner) to the solidifying melt to catalyze heterogeneous nucleation, for instance, Al-Ti-B master alloy for Al-based alloys and Mg-Zr master alloy for Mg-based alloys. A downside of such chemical approach to grain refinement is that the residual of the grain refiner can cause severe degradation to the mechanical properties and corrosion resistance of the final products produced by thermo-mechanical processing. In contrast to chemical grain refining, physical grain refining is achieved through the following mechanisms. The first one is enhanced absolute nucleation rate due to the increased physical undercooling under intensive forced convection (Figure 7), because nucleation rate is proportional to the undercooling at the nucleation temperature. The second one

208

is the increased nuclei survival rate, due to the uniform temperature field and fast heat extraction provided by intensive forced convection, as described by the effective volume nucleation scenario in the previous section.

Figure 8. Physical undercooling of pure Zn as a function of shear rate, obtained by intensive shearing in the twin-screw device. The furnace temperature was 500°C and pouring temperature was 475°C. For a given shear rate, liquid Zn was continuously

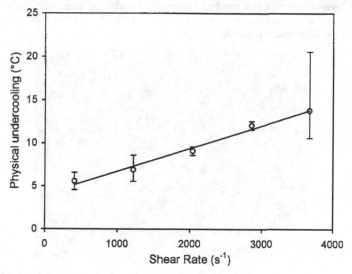

cooled under shear either until the screw stopped or until the melt temperature was constant for 1 minute.

Our experiments have demonstrated that physical grain refining can achieve 35-50μm globule size for most of the engineering alloys, and is more efficient than chemical grain refining, which usually produces a grain size of more than 100μm. In addition, it has also been confirmed that the combination of both chemical and physical grain refining approaches can only achieve very limited extra grain refinement beyond that of physical grain refining alone, 30μm in case of Mg-alloys.

Nucleation control – Physical approach vs. chemical approach

Figure 7 shows that nucleation rate can be enhanced by increasing the physical undercooling (ΔT_P) through increasing Δr by application of intensive forced convection. For simplicity, the critical cluster radii for nucleation $r*$ is assumed to be constant for a given alloy system, Δr can be increased by two totally different ways; one is to raise the equilibrium $r(T)$ curve by selection of suitable alloying elements, this is called chemical approach, while the other one is to increase the intensity of forced convection to push the $r(T)$ curve down, this is called physical approach. It is obvious that the most effective way for nucleation control would be the combination of both physical and chemical approaches. With selection of appropriate alloying elements for larger equilibrium atom clusters in the liquid state, application of intensive forced convection can produce nanocrystalline materials through enhanced nucleation rate, or even amorphous materials due to the sluggish kinetics for crystallization at low temperature. Such combination of both physical and chemical approaches to nucleation control has been demonstrated by our recent experiments.

Figure 9 shows the microstructure of $Mg_{70}Cu_{7.5}Ni_{7.5}Zn_5Y_{10}$ alloy produced by intensive shearing of the liquid using the high shear twin-screw device. The primary phase (α-Mg in an amorphous matrix) has been refined to a few hundred nanometers by physical grain refining. Addition of 5at.% Ag to this alloy produces a fully amorphous material by intensive melt shearing, as confirmed by the XRD results presented in Figure 10.

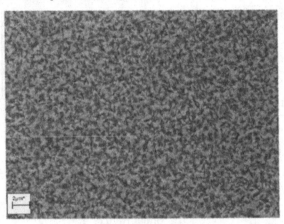

Figure 9. Microstructure of $Mg_{70}Cu_{7.5}Ni_{7.5}Zn_5Y_{10}$ alloy produced by intensive shearing of the liquid alloy using the high shear twin-screw device.

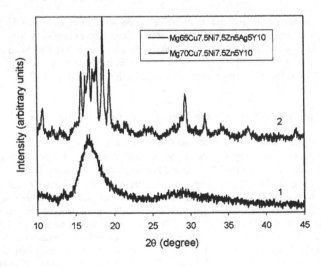

Figure 10. XRD traces of $Mg_{65}Cu_{7.5}Ni_{7.5}Zn_5Ag_5Y_{10}$ (labeled as 1) and $Mg_{70}Cu_{7.5}Ni_{7.5}Zn_5Y_{10}$ (labeled as 2) alloys produced by intensive shearing of the liquid alloy using the high shear twin-screw device.

Summary

(1) The twin-screw device provides a powerful tool for research on solidification under intensive forced convection. Solidification can be studied under unique conditions, such as uniform temperature field, homogeneous composition field, well-dispersed heterogeneous nucleation agents and absence of recalescence. This will eventually push the boundary of solidification from the current static conditions to dynamic conditions.
(2) Under intensive forced convection nucleation becomes continuous, which is in contrast to the conventional "Big Bang" nucleation mechanism, where nucleation is a one off event.
(3) Under intensive forced convection nucleation takes place throughout the entire volume of the liquid, and all the nuclei can survive and contribute to the final solidified microstructure. This is termed 'effective volume nucleation'.
(4) Solidification under intensive forced convection generates nucleation-controlled microstructures with fine and uniform microstructural features, while solidification under static conditions generates growth-controlled microstructures with 3 characteristic zones.
(5) Under intensive forced convection, dendritic growth is inherently supressed, and spherical growth becomes the favourable growth mechanism.
(6) Physical undercooling can be achieved by application of intensive forced convection during solidification, and it increases with the increase of shear rate.
(7) Physical undercooling can result in significant microstructural refinement.

Acknowledgements

The authors would like to thank Dr Y. Wang for his contribution to this work and Professor J. D. Hunt for useful discussions. The financial support from EPSRC (UK) is also acknowledged with gratitude.

References

[1]   W.J. Boettinger, S.R. Coriell, A.L. Greer, A. Karma, W. Kurz, M. Rappaz, R. Trivedi, *Acta Mater.* 48 (2000) 43.
[2]   B. Cantor, *Phil. Trans. R. Soc. Lond. A*, 361 (2003) 409.
[3]   D.M. Stefanescu, *Science and Engineering of Casting Solidification*, Kluwer Academic/Plenum Publishers, New York, 2002.
[4]   Z. Fan, *Mat. Sci. Eng. A*, 413-414 (2005) 72.
[5]   S. Ji, Z. Fan, M.J. Bevis, *Mat. Sci. Eng. A*, 299 (2001) 210.
[6]   Z. Fan, and G. Liu, *Acta Mater.* 53 (2005) 4345.
[7]   D. Turnbull, *Progr. in Met. Phys.* 4 (1953) 333.
[8]   B. Chalmers, *Principles of Solidification*, New York, John Wiley &Sons, Inc. 1995.
[9]   Z. Fan, G. Liu, and M. Hitchcock, *Mat. Sci. Eng. A*, 413-414 (2005) 229.
[10] W.W. Mullins, R.F. Sekerka, *J. Appl. Phys.* 34 (1963) 323.
[11] Z. Fan, X. Fang, and S. Ji, *Mater. Sci. Eng. A*, 412 (2005) 298.
[12] Z. Fan, S. Ji, G. Liu, *Mater. Sci. Forum*, 488-489 (2005) 405.
[13] S.M. Zhang, Z. Fan, Z. Zhen, *Mater. Sci. Tech.* 22 (2006) in press.
[14] A. Das, S. Ji, Z. Fan, *Acta Mater.* 50 (2002) 4571.

[15] Z. Fan, G. Liu, Y. Wang, *J. Mater. Sci.* 41 (2006) in press.
[16] S. Ji, Z. Fan, Met. *Mater. Trans.* 33A (2002), 3511-3520.
[17] V. Varsani, Z. Fan, *Proc. ICSSP-III*, MNIT Jaipur, 19-23 November 2006.

Materials Processing under the Influence of External Fields
Edited by Qingyou Han, Gerard Ludtka, Qijie Zhai
TMS (The Minerals, Metals & Materials Society), 2007

MONTE CARLO SIMULATION OF GROWTH OF SOLID UNDER FORCED FLUID FLOW

A. Das and Z. Fan

BCAST (Brunel Centre for Advanced Solidification Technology),
Brunel University, Uxbridge, Middlesex, UB8 3PH, UK

Keywords: Solidification, Monte Carlo Simulation, Convection, Growth

Abstract

Solidification under intense convection appears to deviate from that under quiescent condition. Most important effect is increased particle refinement and non-dendritic morphology. Despite such interesting observation and the potential for commercial exploitation, solidification under forced convection is relatively less understood. A detailed Monte Carlo simulation has been performed to understand microstructure formation under intense forced convection considering simultaneous atomic transportation in the bulk liquid, atom attachment kinetics at the solid-liquid interface and capillary effect. Evolution of solid morphology and the growth kinetics is studied under both static (purely diffusive) and melt flow (laminar and turbulent flow) conditions. It is clearly indicated that morphological evolution under intense convection is purely a growth phenomena resulting from the effect of forced fluid flow on the diffusion layer geometry around the growing solid. While increased intensity of the forced convection promotes microstructural coarsening, the nature of the flow appears to be more influential on the evolution of the solid morphology.

Introduction

Microstructure formation during solidification is one of the most investigated areas of materials science due to the obvious implication on mechanical properties. In addition, the academic curiosity of understanding the physical phenomena underlying the multitude of patterns observed in solidification and their stability has contributed to significant research activity in this tremendously important area [1, 2]. The advent of powerful computers and development of new computer simulation techniques has led to a resurgence of interest in solidification research and new insight into the evolution of microstructures have been obtained in recent years [1, 3]. One of the recent developments in solidification is the evolution of semisolid processing, in particular, utilising direct or passive stirring techniques promoting forced convection of much higher intensity compared to buoyancy or temperature driven natural convection during solidification [4]. The microstructural evolution under these conditions show non-dendritic morphology and significant refinement of grain size [5].

In spite of the large volume of research conducted on solidification microstructure evolution, theoretical understanding of the solidification process in stirred melt has not progressed much. Vogel and Cantor concluded from a stability analysis that shearing destabilizes the solid-liquid interface and promotes dendritic growth [6] much in contrast to the experimental observation. Accordingly, Doherty and coworkers have proposed a dendrite fragmentation mechanism due to the bending of dendrite arms from shearing followed by fragmentation due to liquid infiltration [7]. Hellawell, however, argued that shear forces are confined within the elastic limit, and therefore, dendrite fragmentation by a mechanical force is unlikely to occur [8]. Moreover, from 3-dimensional microstructural observation in sheared melts it has been suggested that the dendrite fragmentation mechanism might not be as influential as believed [9]. Furthermore, the observation of rosette morphology under lower intensity shearing has to be explained. Through a coupled free boundary and cellular automaton model Mullis proposed that dendrite bending could give rise to rosette formation [10]. Molenaar *et al.* suggested that rosette morphology evolves due to cellular growth of solidification structures under shear but ignores the formation of spherical particles observed at high shear rate [11].

It is now being realized that the observation of spherical growth under intense convection could be a growth phenomena rather than fragmentation of dendrite [12, 13]. Several recent computer simulation also point to such a possibility [14,15]. The present work, therefore, attempts a generalized explanation on the evolution of different solidification structures under different melt stirring conditions through a Monte Carlo simulation.

The model

The present model takes into account simultaneous atomic movement in the melt to simulate diffusion or convection process; atom attachment kinetics at the solid-liquid interface resulting from the bond formation with the solid phase and the associated changes in the volume and surface energy terms; and surface diffusion of the solidified atoms under the action of capillary forces. To represent a binary alloy melt, a two-dimensional square lattice consisting of 400 x 400 points is filled up sequentially with an *A* or *B* atom according to the average concentration of the alloy. As nucleation is not addressed in the model, solidification is seeded from two rows of solid atom at the top horizontal edge of the lattices or from an isolated solid atom at the center of the lattice depending on whether directional growth from substrate or equiaxed growth in the bulk of melt being considered. As the lattice is completely filled and devoid of vacancies, an atomic exchange mechanism is used to address diffusive or convective motion in the liquid. A random lattice position is chosen at first; if a liquid atom with at least one nearest or next-nearest solid neighbor is found, solidification is attempted as described later. Otherwise, the liquid atom is allowed to exchange position with any of its four neighbors with equal probability (pure diffusive motion) or with a higher probability of exchanging position in a pre-determined direction (convective motion). Allowing the liquid atom to exchange position with another liquid atom more than one lattice spacing away can increase the intensity of convection. Each atom in the liquid experiences on an average one attempted exchange in a single time step. However, as the sites for such potential jumps are chosen at random, within a particular time step some atoms may attempt more than one exchange whereas some atoms may not attempt any jump at all.

Solidification probability of an atom is calculated considering the total Gibbs energy change (ΔG_T) incorporating the changes in volume and surface energy and is given by (say, for a solidifying *A* atom),

$$\Delta G_T = \Delta \mu_A / N_V + (n_{A-A} + \phi n^n_{A-A})E_{A-A} + (n_{A-B} + \phi n^n_{A-B})E_{A-B} + (n_{After} + \phi n^n_{After})E \quad [1]$$

where, $\Delta \mu$ is the change in chemical potential, N_V the Avogadro number, n_{A-A} and n_{A-B} the numbers of nearest *A-A* and *A-B* bonds formed and n_{After} represents the total number of unsatisfied nearest neighbor bonds created at the solid-liquid interface upon solidification of the atom. Superscript *n* to these terms refers to the numbers of corresponding next-nearest neighbor bonds. E_{A-A}, E_{B-B}, and E_{A-B} are the energy values associated with the nearest neighbor solid bonds, and ϕ is the weighing factor for the corresponding energy terms for the next-nearest neighbor solid bonds. E is the energy of an unsatisfied bond at the solid-liquid interface which, for simplicity is taken equal for solid *A* or *B* atoms at the interface. A similar expression can be derived for a solidifying *B* atom. The probability of solidification of the liquid atom *W* ('sticking coefficient') is defined as,

$$W = \frac{1}{\tau_s} \frac{\exp(-\Delta G_T / kT)}{1 + \exp(-\Delta G_T / kT)} \quad [2]$$

where *k* is the Boltzmann constant and τ_s is some arbitrary factor, usually taken as unity.

214

An atom is adsorbed at the surface if a pseudo-random number smaller than W is generated. The possibility for this atom to return back to the liquid is considered and if remelting does not occur then the atom is considered to have solidified. After each successful solidification attempt a surface rearrangement trial is performed by allowing the solidified atom to travel to an energetically more favorable position at the solid-liquid interface within a square area of edge length $(2L+1)\Delta x$ with the solidified atom at the center. With L as an integer and Δx as the lattice parameter, the surface diffusion length SDL is defined as $L\Delta x$. A surface rearrangement can be viewed upon as a melting event at the position considered and a resolidification event at a new position of the solid-liquid interface and essentially takes into account the effect of capillary by keeping the curvature information at the interface. The transition probability (W_{SR}) for surface rearrangement to all probable sites are calculated following a procedure similar to the case of solidification and the atom is moved to the most probable position only if W_{SR} is larger than ½. If many of these positions are equally probable

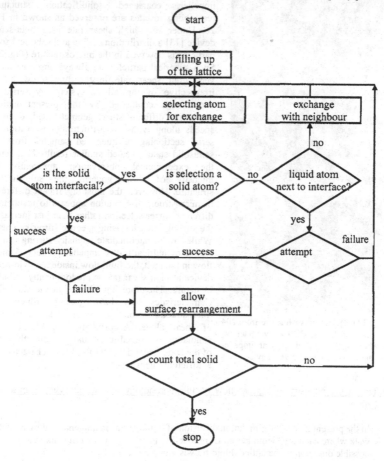

Fig. 1. A flow chart illustrating the important computational steps involved in the present simulation of binary alloy solidification.

then one of them is chosen at random. A flowchart of the simulation is presented in Fig. 1 and the detailed description of the model can be found elsewhere [16,17].

Solidification microstructures formed under different shearing intensity

The morphological evolution of solid under different nature of convection is shown in Fig. 2

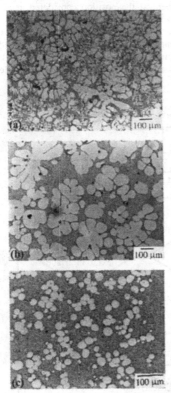

Fig. 2 Quenched microstructure from 204 °C in a Sn-15 wt.% Pb alloy under (a) unsheared, (b) sheared using an impeller (c) sheared in the twin-screw device.

from a Sn-15 wt.% Pb alloy quenched from 204 °C. The light particles represent the solid formed prior to quenching and the fine matrix is the quenched in liquid. Without any forced convection, the solid forms predominantly in a dendritic fashion as shown in Fig. 2a. At low and intermediate shear using an impeller, coarsened solidification structures resembling rosettes are observed as shown in Fig. 2b. Under very high shear rate in a twin-screw device [13] a distribution of fine and spherical solid particles is observed in the microstructure (Fig. 2c). Such spherical particles are always observed under very high shear rate in the twin-screw device irrespective of the alloys system. A series of experiments conducted by the present authors employing different stirrer geometry and rotational speeds along with microstructural observation by serial sectioning of quenched samples from the semisolid state revealed similar results [12]. From the experimental observation apparently the intensity of shear governs the solidification structure. However, the most significant difference in microstructural evolution appears to result from different stirrer selection rather than just increasing the stirring intensity using specific stirrer geometry. While low to intermediate intensity stirring using a simple cylindrical rod or impeller produces laminar flow in the melt, the fluid flow inside the twin-screw device at high shear rate is predominantly turbulent in nature. Therefore, the change in the fluid flow characteristics may have greater influence on microstructure evolution compared to the intensity of shear alone. Accordingly, the Monte Carlo simulation is utilized to evaluate the growth pattern of the solid under different fluid flow characteristics at different intensities.

Effect of the intensity of laminar flow on the morphology of solid growing from a fixed substrate

In the present stochastic simulation diffusion of liquid atoms is implemented by random walk where each melt atom has equal probability to perform a random walk in any possible direction in the lattice. Fluid flow is implementation by allowing a high probability for atomic movement in a specific direction compared to the others; the intensity of flow is determined by how far the atoms can travel in a single jump. The nature of fluid flow is characterized by the direction of gross motion of the liquid atoms relative to the growth direction of the solid rather than trying to simulate liquid motion in a laminar or turbulent motion, which is difficult to implement. This is a reasonable simplification as the growth of the solid is controlled by the interaction of liquid atoms with the diffusion boundary around the liquid-solid interface rather than the exact nature of atomic movement throughout the liquid. A pure shear flow or laminar flow can be visualized as one where mass transport is diffusive inside the diffusion boundary layer (around the particle) and convective outside. Increased intensity of convection is expected to decrease the boundary layer thickness but unlikely to assist forced liquid penetration into the interdendritic region of dendrites growing from a substrate. A turbulent flow, on the other hand, can be visualized as an ensemble of eddies that can destabilize the diffusion boundary layer promoting liquid penetration into the interdendritic regions of the growing solid.

Solidification structures simulated for a binary alloy (C_0 = 0.7) under different intensity of laminar flow conditions are shown in Fig. 3. A solid atom is represented by a red pixel while solvent (A) and solute (B) atoms in the liquid are represented by white and green pixels,

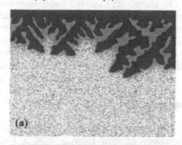

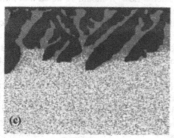

Fig. 3 Effect of (a) diffusive, and laminar flow with atomic movement of (b) 1 lattice spacing and (c) 3 lattice spacing (in a time step) perpendicular to the growth direction of solid on the morphology of solid growing from substrate. Parameters used for the simulation are C_0 = 0.7, $\Delta\mu_A/N_v kT$ = -10, $\Delta\mu_B$ = $5 N_v kT$, E = $2kT$, E_{A-A} = E_{B-B} = $-2kT$, E_{A-B} = -0.1kT, ϕ = 0.1, SDL = 3.

respectively, in the simulated microstructures. Intensity of green pixels represents the local solute concentration in the liquid. As the present model does not address nucleation, solidification on a fixed substrate (say, mould wall) is seeded from two rows of solid atoms at the top edge of the lattice. Values of the simulation parameters are reported in the caption of the Fig. Morphological development under pure diffusive motion, where the liquid atoms have equal probability of exchanging position with any of its neighbor, is shown in Fig. 3a. As expected in a pure diffusive growth, the structure is dendritic with well-developed primary and secondary arms growing along the <1,1> direction in the lattice for energetic reasons. The next-nearest neighbor interaction in these simulations is relatively small (ϕ = 0.1). Therefore, the solidifying atom attempts to maximize the number of satisfied nearest neighbor bonds with the solid (and minimize the number of nearest neighbor broken bonds at the solid-liquid interface). Thus, solidification at {1,1} interfaces is favored because those interfaces have more kinks available than the flat {1,0} interfaces and growth is maximum in the <1,1> directions.

When the liquid atoms are assigned a directional motion perpendicular to the growth direction of the solid to simulate a laminar flow condition, a clear destabilization of the solid-liquid interface has occurred. Dendritic solidification is favored as evidenced from the reduced thickness of the initial solid layer prior to the beginning of dendritic growth in Fig. 3b compared to Fig. 3a. This result is in essential agreement with the theoretical stability analysis of Vogel and Cantor [6]. When the intensity of the laminar flow is increased by allowing the liquid atoms to move three lattice spacings instead of one in a time step, a distinct coarsening of the microstructure is observed although the early destabilization of the solid-liquid interface is still prevalent (Fig. 3c). Coarsening of the dendrites is clearly evident by the thickening of the primary arms, suppression of secondary and tertiary arm formation and the increased smoothness observed over the entire length of the interface.

The morphological evolution can be explained as follows. A lateral motion in the liquid (from left to right) in Fig 2b in essence establishes a laminar type flow in the liquid over a random diffusive motion where no interdendritic liquid penetration occurs but mass transport in the rest of the liquid takes place by convection. Growth is favored at the tips of the protruding instabilities growing in the liquid where the supply of solvent atoms is relatively abundant due to convective mixing in the liquid (compared to purely diffusive conditions). On the other hand, solute rejected in the interdendritic regions (lagging behind the growth front) cannot be transported away and strong growth retardation occurs in these locations. Therefore, a forced laminar flow as in the case of Fig 2b eventually promotes dendritic growth for solid growing from a fixed substrate. The coarsening observed under increased intensity of the laminar flow appears to result from increased liquid penetration in the interdendritic region especially at the tip of the dendrites where the rejected solute is distributed faster due to greater atomic motion. The results suggest a compromise between the availability of solvent atoms and the removal of the solute diffusion layer at the solid-liquid interface.

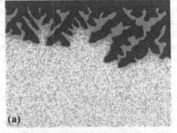
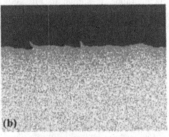
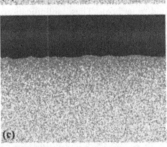

Fig. 4 Effect of (a) diffusive, and turbulent flow with atomic movement of (b) 1 lattice spacing and (c) 3 lattice spacing (in a time step) perpendicular to the growth direction of solid on the morphology of solid growing from substrate. Parameters used for the simulation are $C_0 = 0.7$, $\Delta\mu_A/N_VkT = -10$, $\Delta\mu_B = 5N_VkT$, $E = 2kT$, $E_{A-A} = E_{B-B} = -2kT$, $E_{A-B} = -0.1kT$, $\phi = 0.1$, $SDL = 3$.

Effect of the intensity of turbulent fluid flow on the morphology of solid growing from a fixed substrate

It appears from the previous section that the liquid penetration of the solutal boundary layer promotes coarsening of the microstructure but laminar flow can't produce a stable planar interface growth due to its inability to much reduce or eliminate the formation of the solutal boundary layer. Especially the small pockets of solute rich liquid formed in the interdendritic region are difficult to destroy through such a flow. It should be noted that unlike in pure shear flow, a turbulent flow can be visualized as an ensemble of eddies that can destabilize the diffusion boundary layer promoting liquid penetration into the interdendritic regions of the growing solid. Figure 4 illustrates the morphological evolution of solid growing from the top horizontal edge of the lattice experiencing a liquid motion along the vertical direction (parallel to the growth direction of solid) such that liquid penetrates into the interdendritic regions at all times.

Figure 4a shows dendritic growth under a purely diffusive motion of liquid atoms. By implementing an overall directionality in the motion of the liquid, from the bottom to the top, the dendritic growth of the solid is replaced by more compact growth morphology (Fig. 4b).

218

There is an indication of small instability forming locally at the interface but they appear to die out eventually. On increasing the intensity of fluid motion by three atomic distances, as opposed to one in a time step, the solid liquid interface clearly grows in a stable planar morphology at all times.

A vertical motion of the liquid atoms in the present case ensures constant supply of solvent atoms evenly over the entire interface and the rejected solute is quickly transported away producing a uniformly mixed melt. As at every point of the growing solid-liquid interface the melt concentration is relatively uniform with abundant supply of solvent atoms, perturbation formation at the interface is prevented. As a result, the solid grows more or less with a flat solid-liquid interface giving rise to compact structures. Therefore, the nature of the flow appears to be more effective in governing the morphology of solid forming under forced convection rather than the intensity of the convection. While laminar flow destabilizes the interface promoting dendrite formation, turbulent flow seems to promote interface stability and compact growth morphology. Increased intensity of the flow, however, promotes coarsening effects on the microstructure.

Fig. 5 Growth morphology of an isolated rotating particle suspended in the melt under (a) purely diffusive mass transfer, (b) unidirectional fluid flow from left to right, and (c) liquid impingement from all directions. Parameters used for the simulation are $C_0 = 0.7$, $\Delta\mu_A/N_VkT = -10$, $\Delta\mu_B = 5N_VkT$, $E = 2kT$, $E_{A-A} = E_{B-B} = -2kT$, $E_{A-B} = -0.1kT$, $\phi = 0.1$, $SDL = 3$.

Effect of fluid flow on the equiaxed growth morphology of solid in the melt

So far the simulations have been restricted to solid growing from a fixed substrate under forced fluid motion. However, a great majority of solid forming in the bulk of the melt under forced convection would experience equiaxed growth. This is more appropriate in actual processing, where growing solid particles will be removed from the mould wall under the action of the fluid flow and grow as isolated particles suspended in the melt. Moreover, it has been predicted earlier that the isolated and suspended particles will rotate at a speed of $2\dot{\gamma}$ in a simple shear flow with a shear rate of $\dot{\gamma}$. The particle is effectively rotated by 90° in each time step in a clockwise direction.

Fig. 5a shows equiaxed dendritic growth of a solid particle suspended in the melt under pure diffusive liquid motion with well-developed primary, secondary and tertiary arms. Parameters used for the simulation are reported in the caption of the figure. On implementing a directional motion in the liquid from left to right to simulate a laminar type flow with sufficient intensity the dendritic structure is replaced by a coarse rosette type pattern where the growth of tertiary and secondary arms are suppressed as shown in Fig. 5b. When the liquid atoms are assigned gross directional motion such that the liquid atoms impinge upon the growing solid particle from all directions in a simulated turbulent flow in the melt, a compact solidification morphology evolved (Fig. 5c). The growth of dendritic morphology under pure diffusive liquid motion and compact morphology under liquid impingement on the solid (turbulent motion) can be explained on the basis of the interaction of the liquid motion with the solutal boundary layer formed at the solid-liquid interface as explained in the earlier sections. The rotation of particle under laminar flow promotes increased coarsening leading to rosette morphology during equiaxed growth as compared to that during growth from substrate discussed earlier. It appears that rotation of the particle ensures a periodic change in the liquid flow characteristics around every region of the growing particle. Consequently, every part of the solid-liquid interface alternately experiences a lateral flow promoting dendritic growth (as in the case of growth from fixed substrate) and an impinging flow penetrating the interdendritic regions promoting coarsening resulting in the evolution of rosette morphology.

Discussion

On the basis of the Monte Carlo simulation results an explanation for the microstructural evolution in sheared melts can be presented. Rather than the earlier explanation on the basis of dendritic fragmentation and subsequent coarsening of the fragmented arms, a growth based morphological evolution of solid under melt stirring can now be presented. Under high intensity shearing employing turbulence in the authors group, microstructural observation has always indicated formation of extremely fine and spherical particles irrespective of the alloy system. What is noteworthy is the occurrence of highly globular morphology of the solid from the early stages of solidification, while particle density increases continuously during cooling of the molten alloy in the semisolid region [5, 13]. Dendritic fragmentation cannot account for the highly spherical nature of the particles from the beginning of solidification. A large increase in the particle density can result in special growth of the particles due to diffusional impingement. However, it has been predicted that to maintain spherical shape, the cooling rate have to be substantially lower for low grain density, for e.g, a particle density of 180 mm^{-3} would require a cooling rate below 0.4 K s^{-1} [18]. The observation of spherical growth from the beginning of solidification under turbulence, therefore, cannot be explained on the basis of diffusional impingement and coarsening. From the present investigation, morphological evolution in sheared melts can be entirely explained as a growth phenomena based on the influence of the nature of fluid flow on the geometry of the diffusion zone (diffusion boundary layer) around the growing particle.

Without the application of shear solute transfer takes place by diffusion through the entire volume of liquid (infinite diffusion boundary layer) and the growth structure is purely dendritic as observed in normal casting process (also see Figs. 2a, 3a, 4a). At low and intermediate shear rate a laminar flow is established in the liquid. Under such condition a finite boundary layer exists around the growing solid beyond which the melt becomes homogenized due to the imposed convection. Solute transport inside the finite boundary layer is still diffusive and it has been shown that dendritic growth from a fixed substrate is enhanced under laminar flow (see Fig. 2b). This is in agreement with the earlier prediction by Vogel et al. [6] that convection destabilizes the solid-liquid growth front except that the present analysis predicts destabilization under laminar flow only for directional growth of solid from a substrate. However, the present investigation clearly indicates that increased intensity of the laminar flow promotes coarsening of the microstructure although a compact spherical morphology does not result solely due to coarsening when solid grows from a substrate. Such coarsening of dendritic structures has been observed under low intensity stirring using a cylindrical rod by the present authors [12]. The origin of the coarsened rosette type particles under higher intensity laminar flow conditions can be identified from the present simulation results presented in Fig. 5. For equiaxed growth of suspended solid particles in the bulk of the melt, the coarsening is much effective due to periodic penetration of melt in the interdendritic region resulting in periodic destabilizing and stabilizing effects on the solid-liquid interface. The rosette type morphology therefore results from effective coarsening of the initial dendritic particles (see Fig. 5b). The origin of compact spherical particles observed under turbulence results from the almost complete destruction of the diffusion boundary layer at the solid-liquid interface due to effective liquid penetration over the entire solid-liquid interface. Under such condition solid will grow with a compact morphology as solute will transported away from the interface region discarding any possibility of constitutional undercooling. This explains why spherical solid have been observed from the beginning of solidification under turbulence even when the particle density in the solidifying melt is extremely low.

The growth data obtained from the simulations corresponding to different fluid flow conditions in Figs. 3 are presented in Fig. 6 where the total number of atoms solidified is plotted as a function of Monte Carlo steps. Slope of the plot then defines the average growth rate at any instance (Monte-Carlo step). Similar growth characteristics were observed for identical growth conditions in general. It is clear from Fig. 6 that under laminar flow growth enhancement occurs while under pure turbulent flow progressive growth retardation occurs as compared to pure diffusive flow. As laminar flow enhances dendritic growth, more growth centers (more surface area and kinks) are available for the solidifying atom with the progress of time. In contrast, for pure turbulent flow, the growth rate is high initially due to the faster transport of solvent atoms (aided by convection) towards every region of the solid-liquid interface and gradually retards with the progress of solidification as less growth centers (kinks) are available at the flat solid-liquid interface. The results clearly explain the origin of microstructural refinement of spherical particles formed under turbulent fluid flow conditions. From experimental investigation on solidification under turbulent convection conditions such fast initial growth of solid to a specific size followed by extremely slow growth kinetics have been observed recently and the simulation corroborates the suggestion that the increased particle density is likely to arise from fresh nucleation rather than growth of existing solid particles during continuous cooling [5, 13].

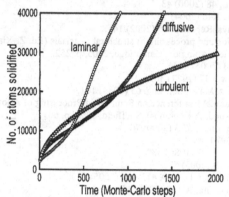

Fig. 6 Number of atoms solidified as a function of time under different fluid flow conditions.

Summary

A Monte-Carlo simulation is presented to investigate the growth morphology of solid under diffusion as well of forced convection, both laminar and turbulent in nature, in a solidifying binary alloy melt. The model accounts for diffusive as well as forced convection of all liquid atoms, attachment kinetics at the solid-liquid interface, and a surface rearrangement process under the effect of capillary force. The results clearly indicate that the morphological evolution of solidification microstructure under forced convection can be solely explained as a growth phenomenon and explains the experimental observations.

Increased intensity of forced convection appears to increase the coarsening of the solid formed. However, the nature of the fluid flow seems to be more influential on determining the growth morphology of solid. A laminar type fluid flow promotes dendritic morphology

221

for solid growing directionally from a fixed substrate, and increased intensity of laminar flow enhances the coarsening of the dendritic structure. For equiaxed growth of solid particles in the bulk of the melt, particle rotation under laminar flow promotes coarsening of dendries to a rosette morphology. A turbulent type fluid flow promotes stable and compact growth morphology of solid for both growth from substrate as well as equiaxed growth. The effect of the nature of fluid flow on the growth morphology results from the effectiveness of the fluid flow in modifying or eliminating the solutal diffusion layer geometry at the growing solid-liquid interface. The simulation result also explains the refined particle morphology observed under turbulence in the melt due to the progressive growth retardation under such conditions.

References

[1]   W. J. Boettinger *et al.*, Acta Mater., 48 (2000) 43.
[2]   U. Hecht et al., Mater. Sci. Eng. R, R46 (2004) 1.
[3]   H. Rafii-Tabar and A. Chirazi, Phys. Reports, 365 (2002) 145.
[4]   A. Das, Z. Fan, in: The deformation and processing of structural materials (Ed. Z. Xiao Guo), Woodhead Publishing Limited, Abington, Cambridge, 2005, p. 252.
[5]   Z. Fan and G. Liu, Acta Mater., 53 (2005) 4345.
[6]   A. Vogel, B. Cantor, J. Crys. Growth 37 (1977) 309.
[7]   R.D. Doherty, H.-I. Lee, E.A. Feest, Mater. Sci. Eng. 65 (1984) 181.
[8]   A. Hellawell, in: Proc. 4th International Conference on Semi-Solid Processing of Alloys and Composites (Eds. D.H. Kirkwood, P. Kapranos), Sheffield, 1996, p. 60.
[9]   B. Niroumand, K. Xia, Mater. Sci. Eng. A283 (2000) 70.
[10] A.M. Mullis, Acta Mater. 47 (1999) 1783.
[11] J.M.M. Molenaar *et al.*, J. Mater. Sci, 21 (1986) 389.
[12] A. Das, Z. Fan, Mater. Sci. Tech., 19 (2003) 573.
[13] Z. Fan, G. Liu, M Hitchcock, Mater. Sci. Eng., A 413-414 (2005) 229.
[14] R.S. Qin, E.R. Wallach, Mater. Sci. Eng, A 357 (2003) 45.
[15] R.S. Qin, Z. Fan, 6th International Conference on Semi-Solid Processing of Alloys and Composites (Eds. G. Chiarmetta *et al.*), Turin, 2000, p. 819
[16] A. Das, E. J. Mittemeijer, Metall. Mater. Trans. A, A31 (2000) 2049.
[17] A. Das, E.J. Mittemeijer, Philos. Mag. A, 81 (2001) 2725.
[18] M.C. Flemings, R.A. Martinez, Solid State Phenomena, 116-117 (2006) 1.

Materials Processing Under the Influence of External Fields

Session V

Steel Production with Microwave Assisted Electric Arc Furnace Technology

Jiann-Yang Hwang, Xiaodi Huang and Shangzhao Shi
Michigan Technological University

Abstract

Microwave assisted electric arc furnace technology is a new steelmaking technology under development at Michigan Technological University. This technology is a revolutionary change from the current technology. It is achieved through the combination of microwave, electric arc and exothermal heating. This technology can produce molten steel directly from a shippable agglomerate (green ball) consisting of iron oxide concentrate, coal, and fluxing agent without the intermediate steps of coking, sintering, Blast Furnace (BF) ironmaking, and Basic Oxygen Furnace (BOF) steelmaking. The unique aspect of this technology utilizes the advantages of rapid volumetric heating, high energy efficiency, and chemical reaction acceleration through the use of microwaves. The viability of the technology lies in the fact that iron ore and carbon are excellent microwave absorbers. This new, simplified process translates into less capital cost, higher productivity, less environmental pollution and treatment cost, higher energy efficiency, and lower production cost.

Keyword: Steel, Microwave, Electric Arc, Coal, Iron, Reduction

Introduction

Steel is a basic material utilized in 80% of the structural components. Construction, equipment, automobiles, airplanes, furniture, appliances, machines, sewage and gas pipes, etc., are just a few examples of industries that use steel. It is a critical material to a country's economy. The economic growth in new developing countries such as China and India has stimulated strong demands in basic materials such as steel. World steel production has exceeded the milestone of a billion tons a year recently. However, the U.S. steel industry has not been able to enjoy much of the global growth. Actually, more than 30 steel companies have claimed bankruptcy in the last five years, including long time giants such as LTV, Bethlehem, National, Weirton, Rouge, etc. Domestic steel production has shrunk from 130 million tons in 1976 to 100 million tons in 2005, while the world steel production increased from 533 million tons in 1976 to over 1 billion tons in 2005 [1]. Currently, the U.S. is the largest importer of steel in the world, importing about 30 million tons per year. The U.S. steel industry is not competitive in the world market. Tariff protection for U.S. steel industry has triggered worldwide conflicts and, as a result, has been abandoned. The shortage of steel and the recent price jumps have left many industries that utilize steel crying for help. The lack of inexpensive, high quality, domestic steel is one of the fundamental problems weakening the manufacturing competitiveness of this country.

The steel produced in the U.S. comes from two types of operations: integrated mills and minimills. Integrated mills utilize a blast furnace to reduce iron oxide to high carbon iron and then remove carbon and other impurities in a basic oxygen furnace to produce steel. Companies such as U.S. Steel, Ispat Inland, and ISG are some examples of integrated mill operations. Steel produced from this route is, in general, high quality. Minimills employ electric arc furnaces (EAF) to remelt steel scrap with DRI (Direct Reduced Iron) for impurity dilution. Steel produced from a minimill is generally lower in quality. Nucor and Gerdau Ameristeel are examples of minimill companies. The trend is that the integrated mill is continuously losing the market share while the minimill grows in the last fifty years. Currently the integrated mills and minimills each account for about 50% of the 100 million tons domestic production.

The primary technical barrier for the U.S. steel industry is how to reduce iron oxide to iron with coal in an efficient, economical and environmentally friendly way. Without this technology, its steel industry will continuously have problems competing in the world marketplace. The blast furnaces in U.S. are old. To build new blast furnaces will cost billions of dollars, which is prohibitive under current economic conditions. Coke is required for the operation of blast furnace. However, environmental regulations have resulted in the shut down of about two thirds of the coke batteries (from 179 in 1979 to 64 (25 coke plants) in 1999) [2]. Relying on imported coke has put many steel mills in jeopardy. The sharp increase in coke price from $80 to $300-400/ton in 2004 has made the problem worse. Minimills traditionally enjoyed an abundant supply of domestic steel scrap. However, this trend has changed because the recent strong demand for scrap internationally has doubled the price of steel scrap. DRI production has also been reduced because its production relies on reformed natural gas. High natural gas prices in recent years have made many DRI plants closed. Sufficient and economic supply of high quality iron to minimills is critical to their future. In addition, minimills can enter into the cold rolling business if they can replace scrap with high quality, low cost iron.

The U.S. has abundant iron ore and coal resources. Developing the technology to produce steel directly from iron ore and coal is among the highest priority as stated in the Department of Energy's 2001 Steel Industry Technology Roadmap.

The Technology

Blast furnace technology for iron production was invented in 1340 A.D.[3] Many incremental improvements have been made and it is the current dominant technology. As shown in Figure 1, sintered iron ore pellets, coke, and lime are charged into a blast furnace. Heated air is blown (blasted) in at high speed (e.g. 500 mph) to partially combust the coke to generate carbon monoxide and heat. Sintered iron ore pellets are reduced to iron hot metal by the carbon monoxide gas at temperatures up to about 3800°F (2100°C). This process usually takes 8 hours. This route wastes a large amount of energy and is environmentally unfriendly. The hot metal is then sent to a basic oxygen furnace (BOF) where pure oxygen is blown into the hot metal to remove the carbon and convert iron into steel. Figure 2 shows the flow sheet of minimill process. Electric arc furnace is very efficient for smelting scrap. However, it requires DRI or other hot metals to dilute the impurities from the scraps [4]. DRI and scrap are both difficult to obtain due to the strong demand and high natural gas price.

To make steel from iron ore and coal fundamentally involves the reaction of iron oxide with reducing gases, such as carbon monoxide and hydrogen, at high temperature and the melting of the reduced iron. Several technical difficulties related to heat and mass transfer are encountered: i) the transfer of heat into the iron oxide pellet (pellet needed for handling purposes), ii) the generation of reducing gases and their diffusion into the pellets, iii) the ratio of carbon monoxide to carbon dioxide, and iv) temperature control. Technologies which can provide the best control of these factors, especially to the center portion of the pellet, will be the choice of the future.

Figure 1. Generalized steel production flow chart in integrated mills.

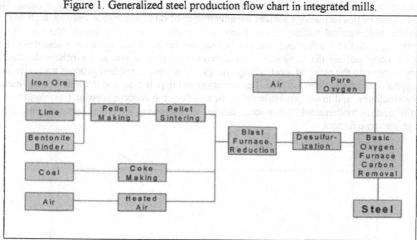

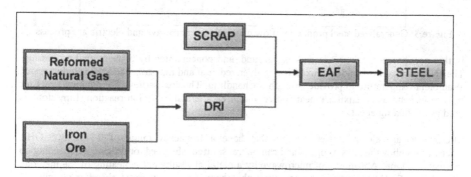

Figure 2. Generalized steel production flow chart in minimills.

The newly invented microwave assisted EAF steelmaking technology is a revolutionary change from the current technology [5, 6]. It is achieved through the combination of microwave, electric arc and exothermal heating (Figure 3). This technology can produce molten steel directly from a shippable agglomerate (green ball) consisting of iron oxide concentrate, coal, and fluxing agent without the intermediate steps of coking, sintering, BF ironmaking, and BOF steelmaking. The unique aspect of this technology utilizes the advantages of rapid volumetric heating, high energy efficiency, and chemical reaction acceleration through the use of microwaves. The viability of the technology lies in the fact that iron ore and carbon are excellent microwave absorbers. The overall concept utilizes the combination of microwaves, electric arc, and exothermal reaction heating to provide the required steelmaking energy. This new, simplified process translates into less capital cost, higher productivity, less environmental pollution and treatment cost, higher energy efficiency, and lower production cost. In addition, the concept remains flexible in that it can still produce molten steel from a scrap charge, thus allowing different feed stocks to be run through the same furnace.

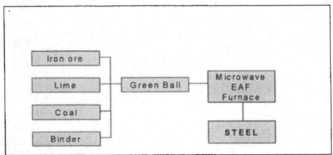

Figure 3. Generalized steel production flow chart of the microwave and electric arc process.

In this new process, iron ore is crushed, ground, and concentrated by conventional processing. The concentrated iron oxide is mixed with pulverized coal and limestone, and then agglomerated at ambient temperature to provide strength for handling. The coal serves as a reducing agent for iron oxides and as an auxiliary heat source via an exothermal oxidation reaction. Limestone is used as the fluxing agent.

Microwaves are electromagnetic waves that have a frequency range of 0.3 to 300 GHz. Microwaves obey the laws of optics and can be transmitted, absorbed, or reflected, depending on the material type. Absorption of microwave by a material results in the heating of the material. Carbon and Fe_3O_4 are excellent microwave absorbers. Fe_2O_3 is a good absorber yet slightly slower than Fe_3O_4, however, they both reach the same final temperature [7, 8, 9, 10,11]. Table 1 list the heating rates of various materials and illustrates the promising utilization of microwave technology for producing steel. Another interesting point noted in Table 1 is the high heating rates of MnO_2 and Ni_2O_3 and the futuristic potential of oxide alloying.

Non-electrically conductive materials normally have poor thermal conductivity and cannot be heated efficiently by an external heating method, especially powdered materials such as an iron oxide concentrate. This is one of the fundamental problems associated with conventional steelmaking. Microwave heating is an internal, volumetric, homogeneous heating method which shows great heating efficiency, when applied to absorber type materials. Heating by microwaves is even more effective on powdered materials, including agglomerated powders. Microwave heating efficiency can be as high as 90-95%.

Table 1. Behavior of Materials Under Microwave Radiation	
Material	**Heating Rate**
Fe_3O_4	20°C/Sec.
Carbon	100°C/Sec.
Fe_2O_3	170°C/Min.
MnO_2	>20°C/Sec.
Ni_2O_3	400°C/Min.

Figure 4 shows the experimental results of iron oxide heating with microwave irradiation. Magnetite (Fe_3O_4) can be heated with microwaves to more than 1000°C (1273K) within one minute. This solved the fundamental heat transfer problem in steelmaking.

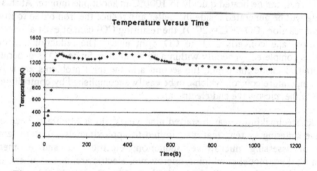

Figure 4. Temperature of magnetite under microwave irradiation.

The equipment for the new technology involved the modification of conventional EAFs with an auxiliary microwave heating system as shown in Figure 5. A port (101) is created in the cover (102) of a conventional EAF to introduce microwaves (103) into the chamber (104) through a waveguide (105). A charge of iron oxide/coal/limestone (ICL) agglomerate (106) is loaded into the chamber. Microwave energy is introduced through the waveguide, where the agglomerates absorb microwave energy and their temperature rises to the point of coal ignition.

Exothermal heat from the carbon/oxidation reaction (107) is generated to further increase temperature. The iron oxide then reacts with the reductant to become directly reduced iron. The EAF electrodes (108) then descend to provide electric arcing energy to the material, producing molten steel and slag. The molten slag and steel are removed by conventional methods utilizing the tilting of the furnace chamber. Thus the furnace can use feed ranging from 100% scrap to 100% iron ore agglomerate. Continuous mode of operation can also occur, with modifications to the chamber.

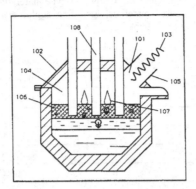

Figure 5. Microwave EAF furnace.

This technology offers the technical solution for producing steel directly from iron oxide and coal as addressed in the technical barrier. First, the volumetric heating provided by microwaves solves the inefficient heat transfer problem associated with pellets. The whole pellet, consisting of iron oxide and coal, can be heated quickly to 1000°C in about one minute. At this temperature, carbon monoxide will be generated inside the pellet and reduce the iron oxide to iron. According to the Boudouard reaction ($CO_2 + C = 2CO$), the resulting CO_2 cannot exist above 1000°C in the presence of carbon, and converts back to CO right away. The reduction reaction will then continue until it is completed, generally within a few minutes. Since CO is generated inside the pellet and maintained inside the pellet, the second and the third problems are solved. The last problem is the temperature control, which can be easily accomplished by controlling the intensity and distribution of the microwave and electric arc.

Microwave direct steelmaking research started nine years ago at Michigan Tech using a single mode microwave sintering system, that was limited to approximately 10 gram samples. After a lengthy effort of assessing microwave, induction, electric arc, and exothermal heating combinations, a working model combining microwave, electric arc, and exothermal heating was identified. This model inspired the design and fabrication of a bench-scale microwave electric arc furnace to handle larger sample sizes of agglomerates, see Figure 6. The furnace is capable of delivering 6KW microwave power (by integrating six 1 KW magnetron from household microwave ovens, mounted on the side) and 12KW electric arc power (obtained from a welding machine, with the electrode shown on the top) to the chamber simultaneously or separately. Molten steel can be produced in the furnace with five minutes of microwave, followed with 5-10 minutes of electric arc, yielding about 500g of steel. Figure 7 is a photo of the steel product. The total operation time is much shorter than the time required for conventional steelmaking from iron ore concentrate.

Evaluation of the steel products showed that steel containing 0.2% carbon and less than 0.01% sulfur can be produced. Steel yield above 95% can be easily obtained. Analyses of steel produced from some of the experiments are shown in Table 2. The exhaust emission gas from the operation was analyzed to assess its environmental impacts.

Figure 6. Bench Microwave EAF furnace

Table 2. Steels Produced by Series #1 MW/EAF Steelmaking Tests.

Steel Sample	Fe	Si	Al	P	S	C
MV-713-ST	97.34	1.65	1.01	<0.01	0.03	<0.2
MV-714-ST	98.54	0.18	0.19	<0.01	0.14	<0.2
MV-718-ST	98.84	<0.01	<0.01	<0.01	0.19	2.32
MV-719-ST	98.65	0.42	<0.01	<0.01	<0.01	1.34
MV-1025-ST	99.30	0.39	0.32	<0.01	<0.01	<0.2

The theoretical energy consumption of the Microwave EAF process is 6.78 MJ/Kg, as shown in Table 3. The actual energy consumption is in the range of 9.2 to 13.56 MJ/Kg at this stage. The recoverable energy from the operation is not deducted. The overall energy including the mining, shipping, and materials energy is 15.04 to 19.4 MJ/Kg. This is a significant reduction comparing with the conventional integrated mill operation, which consumes 27.92 MJ/Kg of energy. An average of 38% energy reduction is realized. If the mining, shipping, and lime energy is not considered, an average of 48.5% of energy reduction can be obtained.

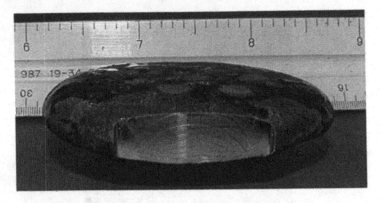

Figure 7. Photograph of Steel product from Furnace in Figure 6.

Conclusions

The opportunity exists now to create a new generation of the steel industry that can be

Table 3. Energy Assessment						
Conventional Steelmaking					Microwave Assisted Direct Steelmaking	
Process	Energy Consumption (MJ/kg)	Process	Energy Consumption (MJ/kg)		Process	Energy Consumption (MJ/kg)
Iron ore processing	2.5	Iron ore processing	2.5		Iron ore processing	2.5
Shipping	1.1	Shipping	0.94		Shipping	0.94
Pelletizing	1.9	Pelletizing	1.9		Pelletizing	1.9
Limestone calcination	1.1	Direct reduction*	14.9		Limestone calcination	0.5
Sintering	1.7	EAF	2.77 (1.3)		MEAF	9.2-13.56 (6.78)
Coke making	1.67					
BF	17.0 (9.8)					
BOF	0.95					
Total	27.92	Total	23.01		Total	15.04-19.4
* Midrex process						

competitive in the world. Strong demands for steel, both domestically and worldwide, especially from developing countries such as China and India, have provided a new platform for the entrance of new technology. This window of opportunity can rejuvenate the steel industry in the US and, consequently, manufacturing and other related industries.

The future users of the new microwave technology can be the existing steel producers as well as newcomers. This technology can be incorporated into existing steel mills. Minimills, which use electric arc furnaces to melt scrap, can modify their furnaces to incorporate the microwave operation. This can help improve the quality of the steel from the impurities in the scrap. The high scrap price due to strong foreign demand has eaten away the profits from minimills and threatens the future supply and existence of minimills. The new technology can provide a low cost iron source to assure the future competitiveness of mini-mills. The new technology can also be part of the integrated mill to provide high quality, low cost iron for the basic oxygen furnace operations. By avoiding the use of coke and sintering pellets, integrated mills will be able to rejuvenate their business by retiring their high cost blast furnace operations.

There will potentially be many newcomers to the steel industry due to this technology. Iron ore companies can initiate their own steel production business because they won't have to rely on metallurgical coke any more. Western coal is abundant, low cost, and readily available. This will provide a new opportunity for those in the industry. Parts manufacturers such as foundries and automobile companies may also have the potential to build small steel producing facilities in their own plants. The new technology will be capable of producing steel at low cost for small operations, and thus, avoiding extra cost for cooling, reheating and transportation of steel from large steel mills to foundries.

Since the technology can potentially save about 50% of the energy currently used to produce steel, its benefit to the environment is tremendous. The steel industry is one of the most energy intensive industries and consumes about 8% of total manufacturing energy. Reduction of this energy by half will reduce energy related pollution such as greenhouse gases and acid rain precursors. In addition, the avoidance of coke utilization is important because coke production is not environmentally friendly.

The price of steel is in the range of $400 to $650 per ton depending on the quality. This new technology has the potential to be very profitable. The base cost of the technology is primarily sensitive to the costs of coal, iron ore and electricity. The prices of these commodities fluctuate much less compared to others such as coke, gas, and oils. The abundance of iron ore and coal in the world will assure the stability of the supply and price of these materials in the future.

References

[1] USGS, Mineral Commodity Summaries: Iron and Steel, 2005.
[2] D.C. Ailor, PRINCIPAL ENVIRONMENTAL ISSUES FACING THE U.S. COKE INDUSTRY IN 1999 AND BEYOND, American Coke and Coal Chemicals Institute, Washington, DC, 1999.
[3] D.T. Llewellyn and R.C. Hudd, Steels: Metallurgy and Applications, 3rd ed., Butterworth-Heinemann, 1999.
[4] A.D. Pelton and C.W. Bale, 1999, Chapter 3, Thermodynamics, pp. 25-41, Direct Reduced Iron, ed. By J. Feinman and D.R. Mac Rae, The Iron & Steel Society.
[5] X. Huang and J.Y. Hwang, "Method for Direct Metal Making by Microwave Energy", U.S. Patent 6,277,168, 2001.
[6] J.Y. Hwang, X. Huang, S. Qu, Y. Wang, S. Shi and G. Caneba, "Iron Oxide Reduction with Conventional and Microwave Heating Under CO and H2 Atmospheres", EPD Congress 2006, edited by S.M. Howard et al, TMS Publications, pp. 219-227.

[7] J.D. Ford, D.C.T., Pei, 1967. High temperature chemical processing via microwave absorption. J. Microwave Power 2(2), 61-64

[8] D.Wong, 1975. Microwave dielectric constants of metal oxides at high temperatures. MSc Thesis. Univ. of Alberta, Canada.

[9] W.R. Tinga, 1988. Microwave dielectric constants of metal oxides, part 1 and part 2. Electromagnetic Energy Reviews 1(5), 2-6 1988, and Electromagnetic Energy Reviews 2 (1), 349-351 1989

[10] McGill, S.L., Walkiewicz, J.W., Smyres, G.A., 1988. The effect of power level on microwave heating of selected chemicals and minerals. In: Sutton, W.H. et al. (Eds.) Mat. Res. Soc. Symp. Proc., Reno, NV, M4.6, Vol.124

[11] J.A .Aguilar and I. Gomez, Microwaves applied to carbothermic reduction of iron ore pellets, Journal of Microwave power and electromagnetic energy, Vol.32, No.2, 1997

Materials Processing under the Influence of External Fields
Edited by Qingyou Han, Gerard Ludtka, Qijie Zhai
TMS (The Minerals, Metals & Materials Society), 2007

STUDY ON FRACTURE OF ALUMINUM ALLOY 2014 CONICAL RING IN HOT RING ROLLING UNDER DIFFERENT TEMPERATURE

Hua Lin[1], Gia Geng-Wei[1], Lan Jian[1]

[1]School of Material Science and Engineering, Wuhan University of Technology
Luo Shi Road 122, Wuhan City, Hubei Province, 430070, China

Keywords: Aluminum alloy 2014, Hot ring rolling, Cracking

Abstract

In this paper, various kinds fracture of aluminum alloy 2014 conical ring during hot ring rolling under different temperature were observed and compared by using SEM, EDS. Mechanisms of different fracture types were analyzed and the best temperature for hot ring rolling is decided. The results showed that the characteristic of fracture surface is mainly dimple fracture below 485 °C. The morphology of various kind of dimple fracture is different. This kind of fracture is resulted from inner micro-cracks that have a close relationship with impurity phase and rolling ring temperature. Fracture surface will transfer into intergranular fracture completely when heating up to 485°C. This kind of fracture is resulted from a lot of impurity phase precipitated at grain boundary, meanwhile overheating structure and many micro-cracks occurred. Based on the results of fracture analysis optimum hot ring rolling temperature of aluminum alloy 2014 is also decided.

Introduction

Ring rolling is an advanced manufacturing technology to product seamless ring parts like bearing rings, train wheels or flank rings and so on through reducing thickness, enlarging diameter, cross-section forming on ring rolling facility. Comparing with forge forming, ring rolling has many characteristics such as saving forging force, saving energy, saving raw materials, high production rate and low costs. So it has been employed in the circle of machining, vehicle, train, boat, metallurgy and airplane, etc[1-3].

Aluminum alloy 2014 is a kind of wrought aluminum alloy with high strength and good thermoplastic property and is generally used to product various parts carrying high loading through hot plastic deformation. But aluminum alloy 2014 has a narrow temperature range for hot deformation, that is 450-350°C, according to previous data[4]. The manufacturing technique for a kind of aluminum alloy 2014 conical ring is changed from machining to hot ring rolling according to its production requirement, as shown in Fig 1. The shape of blank is regular conical ring. Hot ring rolling tests were conducted under different temperature ranging from 380-505°C to obtain optimized hot ring rolling parameters. According to observation of aluminum alloy 2014 conical ring rolled under different temperature, cracks in many large-sized occurred on surface of conical rings. And the size of cracks increased with the increasing of temperature. In

235

this paper, fractography of various fracture surfaces, microstructure and impurity phase of aluminum alloy 2014 conical ring formed during hot ring rolling were observed and compared. The reason of presentation of cracks was analyzed.

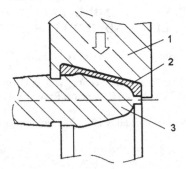

Fig.1. Sketch of hot ring rolling balnk. 1-Driven roll, 2-Conical ring, and 3-Mandrel roller.

Experimental Materials and Setup

The materials for hot ring rolling in this test are commercially-available forged aluminum alloy 2014 club, following by annealing at 400°C and air cooling. The chemical composition of aluminum alloy 2014 is given in Table 1. The blank of aluminum alloy 2014 conical ring is a regular conical ring with 40mm in diameter on top plane and 80mm on bottom plane, 80mm in height and 15mm in thickness. The blank is heated up to desired temperature slowly in resistance furnace and holding time for 30 min.

Table 1. Chemical composition of aluminum alloy 2014 (w)%

Cu	Mg	Si	Mn	Fe	Zn	Ni
4.3	0.6	1.0	0.7	0.6	0.25	0.1

The hot ring rolling tests were executed on the D51-160A lab hot ring rolling mill. The movement of driven roller is manual control mode. The preheating temperature of driven roller and mandrel roller is heat up to about 150°C before rolling. The rolling time of every conical ring is 20-30 sec.

Various conical rings with cracks or the broken conical rings which were rolled at different temperature ranging from 380-505°C were selected. The following test items were carried out:

(1) The microstructure of aluminum alloy 2014 blank was observed by using Pilips Quanta400 SEM.
(2) The fractography of different kind of fracture surface was observed and compared by using SEM and EDS.
(3) The microstructure and defects of aluminum alloy 2014 near the site of cracks were observed by using SEM and EDS. Etchants are keller agent, a acid mixture of HF、HCl、HNO3 and H2O with a ratio of 1:1:2:95. Corrosion time is 20-30 sec.

Experimental Results and Discussion

<u>Microstructure of blank</u>

The microstructure of aluminum alloy 2014 blank after forging and annealing is uniformly distributed equiaxed grains with an average diameter of 40-50 μm, as shown in Fig 2.

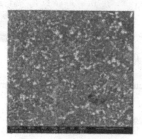

Fig.2 Microstructure of aluminum alloy 2014 blank

<u>Fractography of fracture surface below 400 °C</u>

Cracking of aluminum alloy 2014 conical ring below 400°C often occurred in the edge of top or bottom planes. Cracks are fine and short and the average length is less than 2 mm. Microstructure of such kind of fracture surface is typical equal-axis dimple fracture mixed with large size equiaxed dimples with an average diameter of 10-20μm and small size ones with an average diameter of less than 1μm, as shown in Fig 3a. There are particles in some large size of dimples which contain rich chemical element Si, Cu, Mg according to EDS test results. Microstructure of conical ring is uniformly distributed strengthening phases and impurity phase along grain boundary, as shown in Fig 3b and c.

The impurity phase formed by alloy elements and impurity elements play a role of notch and tend to change into crack sources during hot ring rolling. Generally impurity phase particles distribute along grain boundaries or inner the grains. Stress concentration will take place due to dislocation pile-up around impurity phase particles during plastic deformation. Fine cracks will be produced along the boundary of impurity phase particles and matrix when stress concentration reach to a high level. The fine cracks tend to transform to macro-cracks with the process of hot ring rolling.

Apart from impurity phase particles the second phase particles also contribute to cracks forming. The number of the second phase particles is much more than that of impurity phase particles and average diameter is about 50-100nm[5]. So the role of the second phase particles is contribute to the formation of small size dimples in trans-granular dimple fracture.

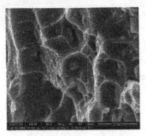

| (a) Fractography of fracture surface | (b) Strengthening phases inner grains | (c) Impurity phases at grain boundary |

Figure. 3 Fractography of fracture surface and microstructure of conical ring below 400℃

Fractography of fracture s urface between 400-470℃

In the case of rolling under the temperature of 400-470℃, cracks often occurred on top or bottom planes and initiated form the site of voids, as shown in Fig 4(a). In stead of equiaxed dimples, most dimples were elongated to ribbon structure mixed with a mount of discontinuous inclusion strips with the length of 10-300μm, as shown in Fig 4(b). The direction of inclusion strips is same as that of metal flow. According to EDS results, inclusion strips contain rich such chemical elements as Cu, Si etc. In local region on the fracture surface the shape of dimples is symmetrical parabola. Average diameter of dimple is about 20-50μm and is larger than that of below 400℃. The number of small dimples is clearly decreased and the height of dimple edge is higher than before.

The microstructure of conical ring rolled at 460℃ is shown in Fig 4(c). Strengthening phases and impurity phases were both elongated and distributed in shape of stripes.

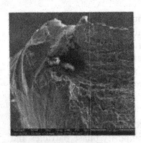
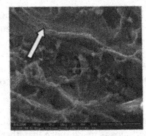

| (a) Site of crack initiation rolled at 460℃ | (b) Fractography of fracture surface | (c) Microstructure of conical ring |

Figure. 4 Fractography of fracture surface and microstructure of conical ring between 400-470□

238

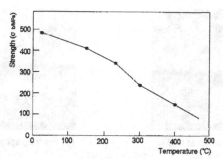

Fig .5. Relationship between temperature and strength of 2014

The coarse impurity phases including (FeMnSi)Al6 and Mn3SiAl12 were broken and arranged in rows in the same direction of rolling. Such kind of coarse impurity is hard to dissolve into α-Al matrix and always play the role of notch in materials. Fine cracks initiate inside or around the impurity phases under the effect of stress during the course of deformation. When the magnitude of stress in the tip of crack exceeds the yield strength of 2014 local cracking in α-Al matrix occurred. So finally banded dimples distributed on the fracture surface of conical ring due to the rows of impurity phases. The amount of strengthening phases dissolved into α-Al matrix become more and more when the temperature is above 400°C. The strength of aluminum alloy 2014 also decreases with the increasing of temperature, as shown in Fig 5[6]. Flow stress of 2014 decreases apparently and plastic flow of material becomes more easy with increasing of temperature. Material flowed under the effect of rolling force and grains were stretched at same time. The precipitates at grain boundary increased and decreased the strength of grain boundary. So in local region on fracture surface equal-axis dimples changed into stretched, parabola dimples that fracture along grain boundary. And meantime small size dimples disappeared gradually.

Fractographyof fracture surface above 490 °C

The number of dimples decreases with the increasing of temperature gradually companied by the transformation of the fracture surface fractograph from dimple fracture to crystal fragile fracture, as shown in Figure 6a, a mixed fracture mode in local region of fracture surface at 490°C. When temperature reach to 495°C fracture mode of aluminum alloy 2014 conical ring will change into complete brittle fracture. Surface of conical ring became dim and dark, and some bubbles were found on surface of conical ring. The morphology of fracture surface were typical inter-granular fracture and all dimples disappeared. There was a mount of micro-cracks along grain boundary. And grain boundary became rough where contain some eutectic balls. Part of grain boundary melted into triple grain boundary among the conjunction of three grains, as shown in Fig 6b. A lot of article-like precipitate was found inside grains, as shown in Fig 6c.

Microstructure of over-heated conical ring is shown in Fig 6d. The size of grain is larger than that of blank and average diameter of grain is 80-120μm. It can be seen that the grains were subject to almost no any plastic deformation before brittle fracture. Among grain boundary there distributed stripe precipitates.

(a) Mixed fracture mode at 490°C

(b) Fractography of fracture surface

(c) Article-like precipitate inside grains

(d) Microstructure of over-heated conical ring

Figure. 6 Fractography of fracture surface and microstructure of conical ring above 490□

Determination of hot ring rolling temperature

2014 is a kind of aluminum alloy with high strength. Because of it's high alloying degree and complex intermetallic compounds low melting eutectic microstructure are easy to segregate at grain boundaries. So hot ring rolling temperature must be below melting point of eutectic microstructure. A little higher temperature than melting point of eutectic microstructure will lead to appearance of over-heated structure.

During hot ring rolling the characteristics of inter-granular fracture become more and more apparent with the increasing of temperature. The dimple is still dominant in fractography at 485 °C and meantime aluminum alloy 2014 can hold a relative good deformation ability. Continuous temperature increasing above 490°C will produce overheated microstructure gradually.

Plasticity of aluminum alloy 2014 begin decline at 490°C when over-heated structure appear. According to the characteristics of fractograph aluminum alloy 2014 could be hot rolled in a temperature range of 470-485°C, that means a improvement of 10-20°C than ordinary forging temperature.

Conclusion

(1) The fractography of aluminum alloy 2014 conical ring below 400°C is mainly transgranular dimple fracture. With increasing of temperature dimple fracture gradually transform to inter-granular fracture. Some over-heated structure begins to form at 490°C and the type of fracture surface is typical over-heated structure above 495°C.

(2) The cracking of aluminum alloy 2014 conical ring below 400°C has a close relationship to rolling parameters, mainly strain rate. Proper rolling parameters can decrease the number of cracks. The cracking occurred above 450°C is mainly due to the precipitates at grain boundary that decrease the strength of grain boundary.

(3) The temperature which is suitable for hot ring-rolling for aluminum alloy 2014 can be increased a little than that of general forging, that means up to 470-485°C.

Reference

1. Hua Lin, Huang XingGao and Zhu ChunDong. *Theory and Technology on Ring-rolling*. (Beijing: Mechnical Industry Press, 2001), 1.
2. Hua Lin and Zhao Zhongzhi, "Ring Rolling Technology and its Application in Automotive Industry," *Automotive Engineering*, 4(1993), 22-25.
3. Hua Lin, Zhao Zhongzhi and Wang Huachang, "Principle and Design Method for Ring Rolling," *Chinese Journal of Mechanical Engineering*, 6(1996), 45-48.
4. Wu Gong, Yao LiangJun and Li ZhengXia. Handbook of Aluminum and Aluminum alloys. (Beijing: Science and Technology Press, 1994).
5. L.F.Mondolfo. Aluminum Alloys: Structure and Propertys. (Beijing: Metallurgical Industry Press, 1988)
6. Xia Gongchen. Handbook of Engineering Material. (Beijing: China Standard Press, 1989)

AUTHOR INDEX

SUBJECT INDEX